Tucholsky Wagner Zola Scott Sydow Freud Schlegel
Turgenev Wallace Fonatne
Twain Walther von der Vogelweide Fouqué Friedrich II. von Preußen
Weber Freiligrath Frey
Fechner Fichte Weiße Rose von Fallersleben Kant Ernst Frommel
Richthofen
Hölderlin
Engels Fielding Eichendorff Tacitus Dumas
Fehrs Faber Flaubert
Eliasberg Ebner Eschenbach
Feuerbach Maximilian I. von Habsburg Fock Eliot Zweig
Ewald Vergil
Goethe London
Elisabeth von Österreich
Mendelssohn Balzac Shakespeare Dostojewski Ganghofer
Lichtenberg Rathenau Doyle Gjellerup
Trackl Stevenson Hambruch
Mommsen Tolstoi Lenz Droste-Hülshoff
Thoma Hanrieder
von Arnim Hägele Humboldt
Dach Verne Hauff
Reuter Rousseau Hagen Hauptmann Gautier
Karrillon Garschin Baudelaire
Defoe Hebbel
Damaschke Descartes
Hegel Kussmaul Herder
Wolfram von Eschenbach Schopenhauer Rilke George
Darwin Dickens Grimm Jerome
Bronner Melville Bebel Proust
Campe Horváth Aristoteles Voltaire Federer Herodot
Bismarck Vigny Barlach Heine
Gengenbach
Storm Casanova Tersteegen Grillparzer Georgy
Chamberlain Lessing Gilm Gryphius
Brentano Langbein
Strachwitz Claudius Schiller Lafontaine Kralik Iffland Sokrates
Bellamy Schilling
Katharina II. von Rußland Gerstäcker Raabe Gibbon Tschechow
Löns Hesse Hoffmann Gogol Wilde Gleim Vulpius
Luther Heym Hofmannsthal Morgenstern
Roth Klee Hölty Goedicke
Heyse Klopstock Kleist
Luxemburg Puschkin Homer Mörike
La Roche Horaz Musil
Machiavelli Kraft Kraus
Navarra Aurel Musset Kierkegaard
Nestroy Marie de France Lamprecht Kind Kirchhoff Hugo Moltke
Laotse Ipsen Liebknecht
Nietzsche Nansen Ringelnatz
Marx Lassalle Gorki Klett Leibniz
von Ossietzky May vom Stein Lawrence Irving
Petalozzi Knigge
Platon Pückler Michelangelo Kock Kafka
Sachs Poe Liebermann Korolenko
de Sade Praetorius Mistral Zetkin

The publishing house tredition has created the series **TREDITION CLASSICS**. It contains classical literature works from over two thousand years. Most of these titles have been out of print and off the bookstore shelves for decades.

The book series is intended to preserve the cultural legacy and to promote the timeless works of classical literature. As a reader of a **TREDITION CLASSICS** book, the reader supports the mission to save many of the amazing works of world literature from oblivion.

The symbol of **TREDITION CLASSICS** is Johannes Gutenberg (1400 – 1468), the inventor of movable type printing.

With the series, tredition intends to make thousands of international literature classics available in printed format again – worldwide.

All books are available at book retailers worldwide in paperback and in hardcover. For more information please visit: www.tredition.com

tredition was established in 2006 by Sandra Latusseck and Soenke Schulz. Based in Hamburg, Germany, tredition offers publishing solutions to authors and publishing houses, combined with worldwide distribution of printed and digital book content. tredition is uniquely positioned to enable authors and publishing houses to create books on their own terms and without conventional manufacturing risks.

For more information please visit: www.tredition.com

The Skilful Cook A Practical Manual of Modern Experience

Mary Harrison

Imprint

This book is part of the TREDITION CLASSICS series.

Author: Mary Harrison
Cover design: toepferschumann, Berlin (Germany)

Publisher: tredition GmbH, Hamburg (Germany)
ISBN: 978-3-8491-7399-9

www.tredition.com
www.tredition.de

TO

HER ROYAL HIGHNESS

THE PRINCESS CHRISTIAN

WHOSE INTEREST IN EVERY GOOD AND USEFUL WORK

HAS SO JUSTLY ENDEARED HER TO ALL

CLASSES OF THIS NATION

THIS WORK

IS

BY HER GRACIOUS PERMISSION

Most Respectfully Dedicated

BY HER

HUMBLE AND OBEDIENT SERVANT

MARY HARRISON

CONTENTS.

1 THE SKILFUL COOK.

INTRODUCTION.

The importance of every woman having a thorough knowledge of domestic economy cannot be too strongly insisted on. The false refinement which, of late years, has considered an acquaintance with domestic matters to be only suitable for servants, has been fraught with the most disastrous consequences. This may seem strong language, but it is not too strong. All sanitary reformers know well enough that it is in the power of many women to prevent very many deaths, and an incalculable amount of misery and vice. Speaking of sanitary reform, the late Canon Kingsley says:— 'Women can do in that work what men cannot. The private correspondence of women, private conversation, private example of ladies, above all of married women, of mothers of families, may do what no legislation can.' And again, in the same speech, delivered on behalf of the Ladies' Sanitary Association, he says:—'Ah! would to God that some man had the pictorial eloquence to put before the mothers of England the mass of preventable agony of mind and body which exists in England, year after year: and would that some man had the logical 2 eloquence to make them understand that it is in their power, in the power of the mothers and wives of the higher classes—I will not say to stop it all, God only knows that—but to stop, as I believe, three-fourths of it.'

This may seem to some, perhaps, too serious an introduction to a cookery book; but it is my earnest wish that my book may not be simply a collection of recipes for cooks to refer to, but a real help to those women who, recognising the importance of good cookery in sanitary reform, are doing their utmost (as I know many are) to acquire that knowledge, and are thereby making the lives of those about them brighter and happier; and are also by their examples doing an amount of good that they themselves scarcely dream of. I have been told more than once by those benevolently interested in the working classes that with instruction to ladies on cookery they had no sympathy, and they seemed to think that it would be better if lessons on the subject were given exclusively to the poor. They forget that the wives of the working men are women who have most of them been domestic servants, and that what they learn in their situations, and what habits they there acquire, they take for *good* or *evil* into their own homes; and in this way an ignorant careless mistress may be doing an infinitude of harm to her sister women in a lower position than herself. On the other hand, a mistress who understands thoroughly the management of a house, by wisely training her servants in habits of order and industry, by teaching them what they do not know and have had no opportunity of learning about hygiene or the laws of health, may be—in fact cannot help being—a blessing indirectly to many homes.

I believe that the working classes must be taught in this way if they are to be taught at all. I have myself, over and over again, tried to benefit my poorer sisters by giving them free lessons on food and cookery; and although 3 I invariably find a few who are very grateful for such instruction, the majority, I imagine, never trouble to put in practice what they have been taught. Their habits have been already formed, and it is not easy for them to alter them. But it is a significant fact that those who do value the lessons are generally respectable hardworking women, who have held good situations under good mistresses.

I have also heard it very ignorantly objected by some that by teaching ladies how to cook, you are taking the bread out of the servants' mouths. This is, indeed, the conclusion of a shallow mind; for with equal justice and good sense, it might be said that the owner of any large business was taking the bread out of his *employés'* mouths because he happened to be acquainted with all the details of his own business, and was able to see that those in his employment attended to their duties properly. But this, I suppose, everyone will admit, that the owner of any business ignorant of the management and details of it, would not unlikely one day find himself without any business to manage. And if this is true with regard to men's businesses, is it not equally so with regard to women's?

I have the greatest sympathy with servants, and would be the last to injure them in any way. A good servant is a treasure: and good work always deserves good wages. But the more a mistress knows of household work herself, the more is she likely to appreciate a servant who honestly and conscientiously performs her duties; and by understanding their difficulties, the more consideration is she likely to show to those in her employ.

But there are some ladies to whom a knowledge of domestic economy ought to be especially invaluable—namely, those whose means are so limited that they cannot afford to engage servants who have had any great experience, and, therefore, who keep only what is called a general servant, a term which often means a woman or girl who will undertake to do everything, but who has only the vaguest notions of how anything should be done. They, poor things, have had no opportunity of learning in the homes from which they came. But it will be well for the poor 'General' if her mistress can teach and train her; for she will then leave her situation with knowledge and habits that will make her a valuable and useful woman, and be of the greatest service to her all her life.

It is, however, quite surprising to see the rough way in which some people allow themselves to be served, and the muddle in which they prefer to live rather than do anything themselves that they consider menial; as if an untidy house, slovenly servants, badly cooked and coarsely served food, are not likely to do much more to lower their self-respect than any amount of so-called drudgery. 'A gentlewoman,' it has been said, 'never lowers herself by doing that which would make her feel less a gentlewoman if left undone.'

How much healthier and happier, too, many girls would be, if, instead of going out in all weathers, day after day, to earn a miserable pittance in any such employment as daily governesses, they would do some of the lighter housework, cooking, &c., at home. By being able to do with one servant instead of two, they would save probably more than they could earn in other ways, besides being much stronger from the exercise thus taken. But too many girls are, unfortunately, imbued with the vulgar notion that work is not genteel. What a Moloch this gentility has been and still is! What a number of human sacrifices are continually placed at its shrine, and what puppets its votaries become! Mr. Smiles says: 'There is a dreadful ambition abroad for being "genteel." We keep up appearances too often at the expense of honesty, and though we may not be rich, yet we must *seem* to be so. We must be "respectable," though only in the meanest sense—in mere vulgar outward show. We have not the courage to go patiently onward in the condition of life in which it has pleased God to call us, but must needs live in some fashionable state to which we ridiculously please to call ourselves; and all to gratify the vanity of that unsubstantial genteel world of which we form a part.'

It would effect a moral revolution if women would only look at matters in the true light. How much crime and misery may be traced to mismanaged unattractive homes! How many deaths to the ignorance of hygiene! How much intemperance to the physical depression caused by badly cooked food! Let us hope that the refinement, falsely so called, which is only another name for vanity, laziness, and selfishness, may soon give way to the true refinement of heart and mind which considers nothing too menial which will benefit others; nothing too common that will add to the happiness of our fellow-creatures.

If we women could earnestly and courageously endeavour to do the duty nearest to us, remembering that all honest work, of whatever kind, has been for ever ennobled by the great Founder of our Faith, so should we be, one in one way and one in another, 'helping to move (to quote Dean Goulburn) the wheels of the great world system whose revolutions are bringing on the kingdom of Christ.' 'To be good and to be useful,' as Canon Kingsley says, 'are the two objects for which we were sent into this world.'

6

HINTS TO YOUNG HOUSEKEEPERS.

She looketh well to the ways of her household.

Proverbs of Solomon.

Take care that you know definitely what sum you can afford to spend on your household expenses, and make it a point of conscience never to exceed it. Market with ready money, if possible; but, if it is more convenient to pay by the month, or quarter, never make that an excuse for letting your bills mount up to double what you can afford to pay. With accounts, carefully kept, it is quite possible to regulate the expenditure to the income.

Never order things at random, but inquire the price of everything before purchasing. Take every pains to know how to judge of the quality of meat, groceries, &c., so that you may not be imposed on. Never be ashamed to say you cannot afford to have this or that. To be poor may be a misfortune, but it is not a fault; and, indeed, to be rich is often a far greater misfortune. The discipline of poverty, and the self-denial it involves, will often strengthen a character which the luxury of riches would enervate.

Cultivate sufficient independence of character to enable you to form your household, and regulate your expenses according to your *own* means, and not according to the income of your neighbours. What does it matter if some may sneer at your thread-bare carpets and frugal fare? The approval of your own conscience is of far more importance than the friendship of the vulgar-minded. Above all things keep your accounts most strictly. Without this you are like a mariner without a compass, or chart, you don't know where you are or what is your position, and you will find yourself, before long, on the rocks of debt and difficulty. Extravagant housekeeping has been the cause of the most serious evils; and, if persisted in, will be sure, in time, to wreck the peace and happiness of yourself and family.

Extravagance is, no doubt, often the result of mere thoughtlessness, but that does not mend matters. There is as much evil wrought

by want of thought as by want of heart. If it is true that there is but
one step between the sublime and the ridiculous, it is equally true
that there is but one step between folly and wickedness. Therefore,
all young housekeepers ought to give earnest attention to the management of their affairs, for certainly in these matters the 'wise
woman buildeth her house, while the foolish plucketh it down with
her hands.'

8

FOOD AND DIET.

The human body is constantly wearing out. With every movement, every breath drawn, there is some waste of its substance. To repair this waste, and, in the case of children, to provide material for their growth, a certain amount of food should be taken daily. The food taken should consist of such qualities as will make flesh and muscle; such as will also keep up the heat of the body, and give force, or the power of movement. These foods must contain a certain quantity of liquid, and the salts necessary to keep the blood pure.

Table of Foods.

Flesh-forming or Nitro-genous.	Heat-giving or Carbonaceous.
Examples — Meat	*Examples* — Butter
Poultry	Suet
Fish	Dripping
Game	And fat of all kinds
Eggs	Sugar in whatever form
Cheese	Starch, which is contained in all vegetables
Flour	
Oatmeal	
Barley	
Rice	
Peas	
Beans	
Lentils	

The foods under the head of flesh-formers, although classed as flesh-formers, are really compound foods. They 9 contain some heat-giving as well as flesh-forming properties.

The heat-giving foods, on the contrary, are all simple foods. Life could not be sustained on any one of them alone, whatever quantity might be taken. These facts are sufficient to show the necessity of a mixed diet. Professor Church says in his lectures on this subject: 'Our food must be palatable, that we may eat it with relish, and get the greatest nourishment from it. The flavour and texture of food, its taste, in fact, stimulates the production of those secretions—such as the saliva and the gastric juice—by the action of which the food is digested or dissolved, and becomes finally a part of the body, or is *assimilated*. As food, then, must be relished it is desirable that it should be varied in character—it should neither be restricted to vegetable products on the one hand, nor to animal substances (including milk and eggs) on the other. By due admixture of these, and by varying, occasionally, the kind of vegetable or meat taken, or the modes of cooking adopted, the necessary constituents of a diet are furnished more cheaply, and at the same time do more efficiently their proper work. Now, if we were to confine ourselves to wheaten bread, we should be obliged to eat in order to obtain our daily supply of albuminoids, or 'flesh-formers,' nearly 4 lb.—an amount that would give us nearly twice as much of the starchy matters which should accompany the albuminoids—or, in other words, it would supply not more than the necessary daily allowance of *nitrogen*, but almost twice the necessary daily allowance of *carbon*. Now animal food is generally richer in albuminoid, or nitrogenous constituents, than vegetable food; so, by mixing lean meat with our bread, we may get a food in which the constituents correspond better to our requirements; for 2 lb. of bread may be substituted by 12 oz. of meat, and yet all the necessary carbon as well as nitrogen be thereby supplied. As such a substitution is often too expensive, owing to the high price of meat—cheese, which is twice as rich in nitrogenous matters (that is flesh-formers) as butchers' meat, may be, and constantly is, employed as a complete diet, and for persons in health, doing hard bodily work, it affords suitable nourishment. Even some vegetable products, rich in nitrogen, as haricot beans, may be used in the same way as meat or cheese, and for the same purpose.' [1]

It is a pity that the value of haricot beans, peas, lentils, and oatmeal is not more generally known. One writer says that there is as much nourishment in 1 lb. of either of these as in 3 lb. of lean meat; and in a lecture on the same subject, another writer states that in three farthings' worth of oatmeal there is as much nourishment as in a mutton chop. These are certainly facts which should be known, especially by people of limited means. Macaroni and semolina are also valuable foods; they are prepared from the most nutritious part of the wheat grain. Rice and maize are deficient in flesh-forming properties, but useful as heat-giving foods; so are, also, tapioca, cornflour, and sago.

Potatoes and fresh vegetables contain but little nourishment. They must not, however, be despised on that account, as they are most valuable additions to our daily diet on account of the potash and other salts which they contain. These vegetables help to keep the blood pure. The anti-scorbutic properties of the potato are so great, that since its introduction into England leprosy is said to have entirely disappeared; neither is scurvy the scourge it was formerly.

The food taken daily should be in proportion to the work done. A labouring man, for example, working hard each day, would require such foods as liver and bacon, steak, bullock's heart, beans, peas, cheese, hard-boiled eggs, &c.; foods, in fact, that would not be too easily digested. Hard work causes the food to be assimilated more readily. A too easily digested fare would cause a constant feeling of hunger. 11 For anyone, on the contrary, leading a sedentary life, the food taken could not be too digestible. In that case, mutton, plainly cooked chicken, soles, milk puddings, and lightly boiled eggs should be the kind of viands chosen.

Children should have plain wholesome fare. Oatmeal and bread are both excellent foods for them. The lime they contain hardens their bones. The bread should be made from seconds flour, which contains more flesh-forming and mineral matter than the whiter and more sifted kinds.

Children should also have plenty of good milk. This is of the greatest importance, especially for the first months of a child's life. Milk is the only perfect food, and contains all that is necessary to sustain healthy life. It is also the only food a child can properly digest, until it cuts its teeth. The improper feeding of children is the great cause of infant mortality. When it becomes advisable to add to milk other foods, they should be nutritious and well cooked. Fine oatmeal or baked flour are, perhaps, the two best. Dr. Fothergill says: 'Children fed on the food of their seniors, or rich cake, and crammed with sweeties, do not as a rule thrive well. They cannot compare favourably with children fed on oatmeal, maize, and milk. Oatmeal is recovering its position as a nursery food, after its temporary banishment. Oatmeal porridge is the food *par excellence* of the infants born north of the Trent, or was, at least, and stalwart people were the results.'

There is no doubt oatmeal is an excellent food, not for children only, but for everyone, especially for those who work hard. It is much to be regretted that it is not more universally used. The English, as a rule, eat too much animal food; and do not give sufficient attention to the proper preparation of vegetables.

Oatmeal water is considered a most strengthening beverage, and is used by men in foundries when beer and fermented liquors would be found too heating.

12 Of alcoholic drinks, Mr. Buckmaster says (echoing the opinion of eminent physiologists): 'Beer, wine, and spirits are never to be regarded as foods. Their popular use is entirely due to their stimulating properties. They contain no nitrogen, and are therefore not flesh-formers, nor can they add anything to the wasting tissues. All stimulants act by increasing, for a time, the vitality of the body; but this activity is always followed by depression in proportion to the previous excitement. Tea and coffee do, to some extent, prevent waste; but their value as foods depends mainly on the sugar and milk taken with them; and their use, *instead of food*, is almost as hurtful as intoxicating drinks. Cocoa differs very much from either tea or coffee, since it is a nutritious liquid food.'

In a lecture on the action of alcohol upon health, Sir Andrew Clark says of health: 'That it is a state which cannot be benefited by alcohol in any degree.' He also states: 'It is capable of proof, beyond all possibility of question, that alcohol, *in ordinary circumstances, not only does not help work, but is a serious hindrance of work.*'

These facts are so important, and ought to be so universally known, that it is to be hoped before long the chemistry of food will occupy the place it should as one of the most necessary branches of everyone's education.

13

THE TABLE.

A properly cooked meal, and a neatly arranged dinner-table, are helps to the happiness and moral progress of the humblest of families.—Buckmaster.

A really capable housekeeper will not be satisfied with good cookery only. She will be careful to have each dish nicely served, however plain it may be. Culture, or the want of it, will be seen at once in the appointment of her table. This remark does not apply to a profusion of glass, silver, or flowers—these are questions of wealth—but to the neatness and order with which a table is laid, and the manner in which the meal is served.

Some people are particularly sensitive to external impressions; and to them a dinner, or any other meal, however costly, served in an untidy room, with table-cloth soiled, silver tarnished, glasses smeared, and above all a slovenly servant, would be enough to give a feeling of depression that would anything but aid digestion.

A great point to be attended to is to have everything perfectly clean and orderly, however old and plain. Clean table-cloths make a wonderful difference to the look of a table; a few flowers also will do much to give it a bright appearance. Servants should be neat in their dress, and quiet in their movements. If only one is kept, that is no reason why she should wait at table in a slovenly dress and with ruffled hair.

The dining-room should be, if possible, a bright room with a good aspect. Heavy, sombre furniture, however 14 fashionable, should be avoided. It is unfortunate that so little attention is paid to the influence of colour; a warm colouring will do much to give a bright look to a room which would otherwise be dull.

The influence of the mental emotions on the digestion is so great that it is important that the conversation at meals should be as cheerful as possible, and no unpleasant subject should be discussed: anything that disturbs the appetite disturbs the digestion also.

With these points carefully attended to—a bright room, neatly-laid table, well-cooked food, and cheerful conversation—dinner, or any other meal, will become what it should be, a refreshment to both mind and body.

15

HOW TO COOK.

Hints to Beginners.

A few hints to beginners on the proper way to set about their work may be, perhaps, of some use; as I know many people get disgusted with cookery at the very outset, and after one attempt, form a resolution never to enter the kitchen again. They have spent the whole morning trying to make a single dish, and that has proved a failure; they have become hot, tired, and irritable, and ill able to bear the laughter their failure has excited. There has been a waste of material to no purpose, and they conclude, therefore, that it is useless for *them* to make any further attempts. At any rate, they determine that they will not try again 'just yet;' and that often means that they do not try again at all. This disappointment and fatigue is generally the result of want of method and forethought. A recipe has been taken into the kitchen to be tried; very probably one half of the terms used in it have not been understood by the would-be cook. She at once begins to make the dish, going to the recipe to look for each article required as she wants to use it. If some of the supplies have run short, she has perhaps to wait in the middle of her operations while she sends to purchase them. Moreover, when the cake, pastry, or whatever it may be, is made, the fire has very likely been forgotten. In this way, even if the dish has been properly prepared, it is spoiled in the cooking.

16 Those, too, who have some knowledge of the art and perhaps, can cook fairly well, will often find the work a great fatigue and toil. They spend double or treble the time they need in the kitchen, just for the want of a little judicious management.

Before trying a recipe read it over, *carefully* notice how a dish is to be cooked, and make up the fire accordingly. If it is pastry, take means to get the oven hot; if a boiled pudding, make a good fire, and put a large saucepan of water on to cook it in before doing any-thing else. When this most important matter is attended to, put all the materials required on the table with the weights and scales; notice what cooking utensils will be required, see that they are all clean and ready for use, and put them near to hand. If, for example, you want to make a cake, proceed in this manner: — Attend first to the fire to get the oven lightly heated, then put out the weights and scales and all necessary materials; put a basin on the table for mixing, two or three cups for breaking eggs in, one or two plates to put the different ingredients on as they are measured, a grater, and anything else that may be required. Then carefully weigh the materials, taking the exact quantities named in the recipe. Prepare them all before mixing any of them. Wash and pick over the currants, and while they are drying, cut up all the candied peel; beat up the eggs, and grease and prepare the cake-tin. The butter should then be rubbed into the flour, and the other dry ingredients should be added. The cake should then be quickly mixed, put into its tin, and placed at once in a hot oven.

If several dishes are to be made, a little thought beforehand will often prevent a very great deal of fatigue and waste of time. Suppose, for example, that you wish to prepare two or three dishes for supper and to make some cakes for tea. You have, perhaps, decided to have a 17 chicken coated with Béchamel sauce, a *gâteau* of apples with whipped cream, a custard pudding, and some rock cakes. Make, the day before, if possible, a list of the articles required for the different dishes, and order what is necessary in good time, so that there may be no delay the next morning. Have the kitchen quite clear from all litters before you begin to work. No one can cook well in a muddle. Then commence operations by making up the fire and putting a saucepan of stock, or water, on to boil for the chicken. Next put the gelatine to soak for the *gâteau*, not forgetting a little in the Béchamel sauce. The longer gelatine soaks, the more quickly it will dissolve. Then slice the apples and put them to stew with the sugar, so that they may be cooking while you are preparing something else. Afterwards truss the chicken; and probably, by the time it is ready, the water or stock in the saucepan will be boiling. Put the chicken into it to simmer gently, noticing the time, so that it may not be over-cooked. Then prepare the ingredients for the rock cakes; mixing them — as they require a quick oven — before the pudding. While they are cooking, prepare the custard; and by the time it is made, the cakes, if the oven is properly hot, will be sufficiently set to admit of the heat being moderated. Now make the Béchamel sauce; strain it and add the dissolved gelatine. Take up the chicken, remove the skewers, place it on a dish, and coat it nicely with the sauce. Then rub the apples through the sieve, and finish making the *gâteau*. By this time the chicken, *gâteau*, and rock cakes are made, and the custard will be cooking. While waiting for the custard, whip the cream for the *gâteau* and put it on a sieve to drain; prepare any decorations you may intend to put on the fowl, and lay them on a plate near to it in the pantry, ready to put on just before serving. Everything will now be ready. With just a little management, even a slow worker would scarcely take a 18 longer time to make these dishes than an hour and a half.

Whatever failures and disappointments you may meet with at first, do not be discouraged. Success is certain if you will only have a little patience and perseverance. Do not be disheartened because you feel very awkward, and because you not unfrequently forget the oven, and let your cakes and pastry burn. Try not to mind the banter of your relations and friends at any possible failure. Many well-meaning efforts to acquire this useful knowledge have been nipped in the bud by the thoughtless, silly way in which some people will laugh at any mistake or blunder. A cake which has caught in baking, or a pudding with the sugar left out, will probably afford them an inexhaustible subject of mirth. Make up your mind, however, not to be discouraged by any of these things. Practice will give nimbleness to your fingers and strength to your memory. As regards any laughter your mistakes may cause, only persevere, and it will not be long before the laugh will be on your side. But keep in mind in any of your attempts that you must be *exact* in all you do. If you try to cook without paying strict attention to weights of the materials to be used and to the other directions, you will deserve to fail. Be very particular in measuring quantities; bear in mind that carelessness in this respect is no mark of a superior cook as some people imagine, but rather of a careless or ignorant one.

As whatever is worth doing at all is worth doing well, bring all your intelligence to bear upon what you take in hand.

19

HOW TO CLEAN STOVES AND COOKING UTENSILS.

Iron Saucepans.

Immerse them in a pan of hot water with soda in it, and wash them thoroughly inside and out, taking care that nothing is left sticking to the bottom of the saucepans. If anything has been burnt in them, boil some strong soda and water in them before washing them, and then rub the bottom of the saucepan with sand until it is quite clean. The sand must be used nearly dry; if too much wetted it loses its power.

The saucepan lids should be thoroughly rinsed and dried.

Enamel Saucepans.

Wash them thoroughly in hot water with soda in it, using soap if necessary. If anything has been burnt in the saucepan, boil strong soda and water in it before cleaning it, and rub it well with sand. Rinse and dry thoroughly.

Anglo-American Saucepans.

Clean like enamel saucepans. They should be kept perfectly clean inside and out.

Tin Saucepans.

Clean these like iron saucepans.

20

Dish Covers and Jelly Moulds.

Wash with soap and water and dry thoroughly. Powder some whiting, and mix with a little cold water; brush the mixture over the covers and moulds; when dry, rub off with a plate brush or soft cloth or leather.

To Clean a Roaster.

Wash the dripping-pan and inside of the roaster with hot water and soda to remove all grease, then rub them with sand until they are quite bright, rinse and dry thoroughly. Clean the outside of the roaster with whiting, used according to directions given for cleaning dish covers.

Hair and Wire Sieves.

Wash these thoroughly with hot water with soda in it, and scrub them quite clean with a sieve-brush. Dry them thoroughly, and keep them in a *dry place*. If this is not done a hair sieve will get mildewed, an iron one rusty, and a copper one will verdigris and become poisonous. Copper-wire sieves should always have especial care.

Paste Boards and Rolling Pins.

Scrub them well with hot water and sand. Do not use soda, as it will make the wood yellow.

Baking Tins.

Wash them in hot water with soda in it, and rub with sand until they are bright; rinse and dry well.

21

To Clean a Close Stove or Open Range.

Scrape out all the ashes and brush up all the dust. Then, with a brush, thoroughly clean the flues. Brush the stove over with liquid blacklead, and when it is dry polish with brushes. Then clean any steel about the stove and the fire-irons and fender with emery-paper; any brass with brick-dust well rubbed on with a leather.

Brush all the dust from the oven, and wipe it round with a cloth wrung out of hot water.

To Clean a Gas Stove.

Wash off any grease that may have been spilled on the stove with a cloth dipped in hot water, and wipe the inside of the stove, taking care to dry it thoroughly. Wash the dripping-pan in hot water with soda in it, and rub it with sand to brighten it. Then wipe it quite dry.

Brush the stove over with liquid blacklead, and polish it with brushes.

Copper Cooking Utensils.

Wash them well in hot water with soda in it; moisten some salt with vinegar, and rub them well with this to remove stains and tarnish. Then wash them quickly with soap and water, and dry them thoroughly; polish them with a little powdered whiting rubbed on with a soft leather.

22

RULES FOR BOILING.

All meat, with the exception of salt meat, should be put into boiling water, and should be well boiled for quite five minutes, in order that the albumen on the outside of the joint may be set. The hardened albumen forms a kind of casing. This casing serves to keep in, as far as possible, the flavour and juices of the meat. When the meat has been boiled sufficiently long to effect this hardening, the kettle should be drawn to one side of the fire. The water should be kept at simmering point until the joint is cooked. The general rule, as regards time required for boiling, is a quarter of an hour for each pound of meat and a quarter of an hour over. But only general rules can be given, as the time will vary according to the nature of the joint to be cooked. A thick piece of meat will necessarily take longer to cook than a thin piece with much bone, although both may be the same weight. Very *fresh* meat will also take longer to cook than that which has been hung.

As soon as the water boils, after the meat is in it, the scum should be carefully removed from time to time, while it is cooking. If the scum be allowed to boil down, it will settle on the joint and discolour it. It is best, however, as a precaution, to wrap the meat in a very clean cloth; this will effectually preserve its colour. Salt meat should be put into lukewarm water, for the purpose of drawing out some of the salt. It should be simmered gently, allowing always twenty minutes to the pound, and twenty minutes over. Salt hardens the fibre of the meat; it, therefore, requires to be cooked for a longer time to make it tender.

23

RULES FOR ROASTING.

To roast successfully, make up a nice clear fire. When once made up, it should be replenished, if necessary, by putting on coal or coke at the back. The live coals should be drawn to the front to prevent smoke. Fasten the joint to the jack. Place the roaster close to the fire for the first ten minutes, so that the heat of the fire may at once harden the albumen, and form a case to keep in the flavour and juices. Afterwards, draw the roaster farther back and cook gradually, basting every ten minutes. The basting keeps the meat from drying up, and gives it a better flavour. The length of time allowed for roasting is the same as for boiling, the rule being a quarter of an hour for each pound, and a quarter of an hour over. For white meat, veal and pork, or solid joints without bone, allow twenty minutes to the pound, and twenty minutes over. These rules, however, cannot always be strictly adhered to, as the size and shape of the joint must be taken into consideration, as well as the weight. Meat that has been frozen will take longer to cook than fresh meat. Meat which has been well hung will take a shorter time than fresh meat. If a jack is not used, the joint should be fastened to a rope of worsted, which should be kept constantly turning.

Gravy, for a joint, may be made according to two methods. The first method is to take the dripping-pan away half an hour before the joint is cooked, then to put a hot dish in its place, and to pour the contents of the pan into a basin. Put the basin into a refrigerator; or, place it on ice. As soon as it is cold, the fat will cake on the top of the gravy, and should be removed very carefully. Make 24 the gravy hot, diluting it with warm water, if necessary, and pour it round the joint.

The other and more usual method of making gravy, is to pour away all the fat from the pan as soon as the joint is cooked; and then pour into the pan a sufficient quantity of hot water, scraping well the brown glaze from the bottom; colour carefully with caramel, or burnt sugar, and pour it *round* the joint, not *over* it. Pouring the gravy over the meat destroys its crispness.

On no account make gravy from stock; stock is quite unsuitable, as the vegetable flavour is, to many persons, disagreeable.

RULES FOR FRYING.

French or Wet Frying.

This is cooking in a large quantity of fat sufficient to cover the articles fried in it. Oil, lard, dripping, or fat rendered down, may be used for this purpose. Oil is considered the best, as it will rise to 600° without burning; other fats get over-heated after 400°, and therefore require greater care in using. Success depends, almost entirely, on getting the fat to the right degree of heat. For ordinary frying, the heat required is 345°. Unless this point is carefully attended to, total failure will be the result. There are signs, however, by which anyone may easily tell when the fat is ready for use. It must be quite still, making no noise; noise, or bubbling, will be caused by the evaporation of moisture, or water in it. The expression, 'boiling lard,' or 'boiling fat,' has been misleading to many inexperienced cooks, who, not unnaturally, imagine that when the fat is bubbling, like boiling water, it is boiling, and, therefore, at the right heat. 25 But boiling *fat* does not bubble. When it has the appearance of boiling water, it is simply due, as already explained, to the presence of water in it, which must pass away by evaporation, before the fat can reach the required heat. When it ceases to make any noise, and is quite still, it should be carefully watched; for very soon a pale blue vapour is seen rising, and then the fat is sufficiently hot. If, from the position of the stove, it is not easy to see this vapour, a piece of bread may be held in the fat as a test; if it begins to turn brown, in about a minute, the fat is ready. It should then be used without delay; since, when once hot enough, it rapidly gets overheated or burnt. Fat is burning when the blue vapour becomes like smoke. Burnt fat has an unpleasant smell, and is apt to give a disagreeable taste to the articles fried in it. With ordinary care fat need not get overheated. Next to oil, fat rendered down (*see* Rendering down Fat), is best for the purpose. If strained after each time of using, and not allowed to burn, it will keep good for months, and may be used for fish, sweets, or savouries, and no taste of anything previously fried in it will be given to the articles cooked. For this kind of frying, a kitchener, or gas stove, is preferable to an open range.

All kinds of rissoles, croquettes, fillets and cutlets of fish, fritters, &c., should be fried in this manner, and should not be darker than a golden brown. It is an advantage to use a frying-basket for all such things as are covered with egg and bread-crumbs; but fritters, or whatever is dipped in batter, should be dropped into the fat, as they become so light that they rise to the top of it. When they are a pale fawn colour on the one side, they should be turned over to the other. Care must be taken to drain everything, after frying, on kitchen paper in order to remove any grease.

26

Dry Frying.

This is frying in a cutlet or frying pan, with a small quantity of fat, and is only suitable for such things as require slow cooking, such as steaks, mutton or veal cutlets, fillets of beef, liver and bacon. Pancakes also are fried in this manner. Success depends, as in French frying, in having the fat rightly heated, taking care that the outside of the meat cooked be sealed up. In this way the juices and flavour will be retained in it. Make, therefore, the frying-pan hot, then put in the fat; and when that is also perfectly hot, put in the meat to be cooked. When each side has been well sealed up, the heat applied must be moderated, so that the cooking may be gradual. The common mistake in this kind of frying is to put the meat into the fat when it is but barely melted; the juices of the meat are thus allowed to escape, and the meat is toughened.

RULES FOR BAKING.

To bake meat successfully, the oven must be well ventilated, otherwise, the joint cooked in this manner will have an unpleasant flavour. The meat should be put on a trivet, which should be placed on a baking-tin. The oven must be very hot when the meat is put into it, and the heat should be kept up for the first quarter of an hour. This is to form the casing already alluded to in the directions for roasting and boiling; the heat of the oven must then be very much moderated, and the joint cooked very gradually, allowing twenty minutes for every pound, and twenty minutes over. The meat should be basted; and the gravy may be made in the same manner as in roasting.

27

RULES FOR GRILLING.

For this method of cookery, a clear fire is essential. The griller is warmed, and the meat fastened in it. It is then hung on the bars of the fireplace, and a dish passed underneath to catch any gravy. An ordinary sized chop, cooked in this way, will take about five minutes on the one side, and three on the other.

RULES FOR BROILING.

This is cooking over the fire on a gridiron. The flavour of broiled meat is usually preferred to that of grilled. Put the gridiron over the fire to heat, and then put the chop, or steak, on it; place the gridiron close to the fire at first, that the heat may rapidly seal up the outside of the meat. When this has been accomplished, lift the gridiron further from the fire, and cook gradually, turning occasionally. A clear fire is essential. Coke is better than coal for broiling, because there is less smoke from it.

28

JOINTS.

Sirloin of Beef.

This is the primest joint, and must be either roasted or baked (see directions). Horse-radish should be served with it. Yorkshire pudding is also liked with roast beef.

Ribs of Beef.

These should be cooked like sirloin, and served with the same accompaniments. A neater looking joint is made by boning and rolling them. The bones can be used for soup.

Aitch Bone, Round, Thick and Thin Flank of Beef.

Those are usually salted and boiled (see directions for boiling salt meat). Serve with carrots and turnips, and yeast, Norfolk, or suet dumplings.

Brisket of Beef.

This should be stewed (see directions for stewed brisket).

Leg of Mutton.

This may be roasted, baked, or boiled. If roasted, it should be served with red-currant jelly; if boiled, with caper sauce. Carrots and turnips are liked with boiled mutton.

29

Shoulder of Mutton.

This may be either roasted or baked. Serve with onion sauce.

Saddle of Mutton.

This may be either roasted or baked. Serve with red-currant jelly.

Neck of Mutton.

This is boiled, and requires long and gentle cooking. Serve with caper sauce.

Fore Quarter of Lamb.

This joint should be roasted or baked. Serve with mint sauce.

Leg of Lamb.

This may be either roasted, baked, or boiled. Serve, if roast, with mint sauce; and if boiled, with *maître d'hôtel* sauce.

Shoulder of Lamb, Saddle of Lamb, Loin of Lamb

All these are either roasted or boiled, and served with mint sauce.

Fillet of Veal.

Stuff it with veal stuffing and make into nice round shape; fasten it securely with string and skewers, and roast or bake it. Serve with cut lemon, and send some boiled ham, pork, or bacon to table with it. Use a pint of thin melted butter, instead of water, for making the gravy.

Breast, Shoulder, and Loin of Veal.

These are all roasted. Thin melted butter is used to make the gravy for them, and cut lemon is served with them.

30

Knuckle of Veal.

This is boiled, and served with one dessertspoonful of chopped parsley added to one pint of melted butter.

Leg of Pork.

This must be roasted or baked, the skin having been previously scored with a knife. Serve it with apple sauce.

Chine of Pork.

Stuff it with pork stuffing (see Forcemeats) and roast it. Serve with apple sauce.

Spare Rib of Pork.

This is roasted, the skin having previously been scored. Serve it with apple sauce.

Hand of Pork.

Soak it for two or three hours before cooking, and boil it. Serve with pease pudding.

Leg of Pork.

This joint is also salted and boiled. It is served with pease pudding.

To Cook a Ham.

Put into lukewarm water, to which has been added one pint of old ale. Simmer it very gently until quite tender. For a ham always allow twenty-five minutes to each pound, and twenty-five minutes over. Let it get cold in the liquor in which it boiled, then remove the rind and carefully cover with raspings.

31

Bacon.

Cook like ham, taking care that it is simmered until perfectly tender. Remove the skin and cover with raspings.

Pickled Pork.

Put it into lukewarm water and simmer gently until tender.

32

POULTRY AND GAME.

Roast Goose.

- *Ingredients*—1 Goose.
- Sage and onion stuffing.
- 1½ oz. of flour.
- 1 onion.
- 1 apple.
- 3 sage leaves.
- ½ lb. of gravy beef.
- 1 quart of water.

Method.—Stuff the goose by placing the sage and onion forcemeat inside it.

Then truss it nicely and roast it from one and a half to two hours.

If it is a large one, two hours; if a small one, one and a half hours.

To make the gravy, simmer the giblets in water for three hours with half a pound of gravy beef cut in pieces, a sliced onion, apple, and three sage leaves, pepper and salt.

Then stir in a thickening made of the flour, and colour the gravy with a little burnt sugar. If liked, a glass of port wine may be added.

Pour a little gravy round the goose, and serve the rest in a tureen.

Apple or tomato sauce should be served with roast goose.

33

Roast Turkey.

- *Ingredients*—1 turkey.
- Some veal forcemeat (omitting the suet).

- 1 lb. of gravy beef.
- 3 pints of water.
- 1 onion.
- 2 oz. of flour.

Method.—Place the forcemeat inside the turkey, and truss it nicely.

Roast it from one and a half to two and a half hours.

Make the gravy by simmering the giblets and beef in the water with the onion for three hours.

Thicken the gravy with the flour, and pour a little round the turkey.

Serve the rest in a tureen.

Place some fried or baked sausages round the turkey, and serve with bread sauce.

Boiled Turkey.

A small turkey is sometimes boiled like a fowl, and served with oyster, celery, or Béchamel sauce.

Roast Duck.

- *Ingredients*—1 duck.
- Some sage and onion stuffing.
- Rather more than 1 pint water.
- 1 oz. of flour.
- 1 onion.
- 1 apple.
- ¼ lb. of gravy beef, or 2 or 3 bones.

Method.—Stuff the duck by placing the forcemeat inside it.

Truss it nicely, and roast it from three-quarters of an hour to an hour, according to its size.

34 Make the gravy by simmering the giblets in the water with the beef or bones, onion, apple, pepper and salt, for two hours.

Thicken it with the flour, and colour carefully with burnt sugar.

Pour a little gravy round the duck, and serve the rest in a tureen.

A glass of port wine may be added to the gravy if liked.

Apple or tomato sauce should be served with roast duck.

Ducklings.

These are cooked and served like ducks, and take from twenty to forty minutes to roast, according to their size.

Roast Hare.

- *Ingredients*—1 hare.
- Some veal forcemeat.
- ½ lb. of gravy beef.
- 1 pint of water.
- 1 onion.
- 1 oz. of flour.
- Pepper and salt.

Method.—Stuff the belly of the hare with the forcemeat, and sew it in.

Truss it nicely, and roast it from one and a quarter to two hours, according to its size, basting it constantly.

To make gravy, cut the beef into small pieces, and simmer in the water, with the onion sliced, for three hours. Thicken it with the flour, and add, if liked, a glass of port wine.

Pour a little gravy round the hare, and serve the remainder in a tureen.

35

Jugged Hare.

- *Ingredients* — 1 hare.
- Some veal forcemeat.
- 2 oz. of butter.
- 1 onion, stuck with 6 cloves.
- 2 glasses of port wine.
- 1½ pint of gravy or stock.
- 1 lemon.

Method. — Dry the hare well and cut it in pieces.

Fry them in the butter.

Then remove them and fry the flour a nice brown.

Pour in the gravy or stock, and stir until it boils.

Then put the stock into an earthenware jar with the hare, onion, thin rind and juice of the lemon, and pepper and salt to taste.

Cover the jar close, and put it into a moderate oven, where it must simmer gently from three to four hours until the hare is quite tender.

Make some balls of veal forcemeat, to which the chopped liver of the hare has been added, and either fry or bake them.

Add them to the jugged hare, and, last of all, pour in the wine.

Serve with red-currant jelly.

Roast Rabbit.

- *Ingredients* — 1 rabbit.

- Some veal forcemeat.
- Some nice gravy (*see* Gravy).

Method.—Fill the belly of the rabbit with the forcemeat, and sew it in.

Truss it nicely, and roast it from three-quarters to one hour, basting constantly.

36 Pour a little gravy round it, and send some to table in a tureen.

Serve with red-currant jelly.

Boiled Rabbit.

- *Ingredients*—1 rabbit.
- Some onion or *maître d'hôtel* sauce.

Method.—Boil the rabbit gently from half an hour to an hour, according to its size and age.

Serve it with onion or *maître d'hôtel* sauce.

Stewed Rabbits.

- *Ingredients*—2 rabbits.
- 4 large onions.
- 3 pints of water.
- 2½ oz. of flour.
- Pepper and salt to taste.

Method.—Cut the rabbits into joints, and slice the onions.

Put them with the water into a large stewpan, and simmer for one hour or more until the rabbits are tender.

Then make a thickening of the flour and stir it in, letting it boil well.

Put the rabbit on a hot dish, and pour the gravy over.

Ragout of Rabbit.

- *Ingredients* — 1 rabbit.
- 1 onion stuck with 6 cloves.
- 2 oz. of butter or dripping.
- 1 oz. of flour.
- 1½ pint of water or stock.
- Pepper and salt to taste.

Method. — Cut the rabbit into neat joints, and fry them in a stew-pan in the butter or dripping.

When brown remove them and fry the flour.

37 Then pour in the water or stock, and stir until it boils.

Put in the pieces of rabbit with the onion, and pepper and salt to taste.

Simmer gently for about one hour or more until quite tender.

Serve the rabbit on a hot dish, and strain the gravy over it.

Roast Pheasant.

- *Ingredients* — 1 pheasant.
- Half a pint of gravy.
- Butter.

Method. — Roast the pheasant nicely for three-quarters of an hour or an hour, according to its size, basting it constantly with butter.

Make a nice gravy for it (*see* Gravy), and serve it with bread sauce and browned crumbs.

Wild Duck.

- *Ingredients* — Wild duck.
- Half a pint of gravy (*see* Gravy).
- Lemon juice.
- Butter.

Method. — Roast the wild duck nicely before a clear fire for thirty or forty-five minutes, basting it constantly with butter.

Sprinkle over it a little cayenne and salt, and a few drops of lemon juice.

Serve the gravy in a tureen.

If liked, a glass of port wine may be poured over the duck.

Partridges.

Partridges should be nicely roasted before a clear fire from twenty-five to thirty minutes.

Serve with a little gravy and bread sauce.

Browned crumbs are also handed with them.

38

Grouse.

Roast these birds before a nice clear fire, basting constantly with butter.

Serve with gravy, bread sauce, and browned crumbs.

Woodcocks and Snipes.

These birds should be nicely trussed but not drawn.

Roast them carefully from twenty to thirty minutes, basting constantly.

Place under them rounds of toasted bread, buttered on each side, to catch the trail as it drops, as this is considered a delicacy.

When cooked, lay the toast on a hot dish, place the birds on it, and pour a little good gravy over.

Boiled Fowl.

Truss nicely and flour the breast slightly.

Fold it in buttered paper, and tie securely with string.

Boil in stock or water, according to the directions given for boiling meat for three-quarters of an hour to one hour and a half, according to its age and size.

Serve with white, egg, or *maître d'hôtel* sauce poured over it.

Roast Fowl.

Truss nicely and roast, according to directions given for roasting meat, for three-quarters of an hour to one hour and a half according to its age and size.

Serve with bread sauce and some gravy (*see* Gravy).

39

Braised Partridges.

- *Ingredients* — A brace of partridges.
- A small piece of carrot, turnip, and onion.
- 2 tomatoes.
- 1 pint of good second stock.
- 1 wineglass of sherry.
- Pepper and salt to taste.

Method.—Truss two partridges as for boiling.

Put at the bottom of a stewpan the vegetables cut in small pieces.

Lay the partridges on the top and pour in the stock and sherry; these should be sufficient to come half way up the partridges.

Cover with buttered paper.

Put the lid on the stewpan and simmer very gently until the partridges are tender.

Then put them on a baking tin in the oven to brown them.

Strain the stock and boil it rapidly down to a glaze.

Serve the partridges with the glaze poured over them.

40

SAVOURY MEAT DISHES.

Stewed Steak.

- *Ingredients* — 1½ lb. of steak.
- 1 piece of carrot, turnip, onion, and celery.
- 1 pint of water.
- 1 oz. of dripping.
- 1 oz. of flour.
- Pepper and salt.

Method. — Cut all the fat from the steak.

Make the dripping hot in a stewpan and fry the steak in it.

Then put in the vegetables, and pour in the water, adding pepper and salt.

Simmer the steak gently from three to four hours, until quite tender.

When quite cooked, remove it from the gravy.

Put it on a hot dish.

Make a thickening of the flour; stir it into the gravy; boil for two minutes, and strain over the steak.

A little mushroom catsup, Harvey, or Worcester sauce may be added if liked.

The fat should previously have been cut into dice, placed on a baking tin, and cooked in the oven.

For serving, put them in the middle of the steak.

41

Stewed Brisket of Beef.

- *Ingredients*—5 lb. of beef.
- 2 carrots.
- 2 onions.
- 2 turnips.
- 1 head of celery.
- 1 sprig of parsley.
- Marjoram and thyme.
- 2 bay leaves.
- 6 cloves.
- 1 dozen peppercorns.
- 3 quarts of water.

Method.—Put the meat into a saucepan with the vegetables and other ingredients, and simmer gently for three hours.

Serve on a hot dish, with some of the liquor for gravy.

The remainder can be made into soup.

If to be eaten cold, remove the bones, and press the beef.

Strain the meat liquor, remove the fat, and boil it down to a glaze.

Brush the meat over with it, giving it as many coats of glaze as necessary.

Stewed Ox-cheek.

- *Ingredients*—1 ox-cheek.
- 1 cowheel.
- 3 or 4 carrots.
- 2 or 3 turnips.
- 3 or 4 onions.
- 1 sprig of parsley, thyme, and marjoram.
- 2 bay leaves.
- 2 quarts of water.

- 4 oz. of flour.

Method.—Wash the ox-cheek and cowheel, and cut them into neat pieces.

42 Put them into the water with the carrots, turnips, and onions, and celery cut in pieces, and the herbs, pepper and salt.

Stew very gently from four to five hours, until the stew is quite tender.

Make a thickening of the flour.

Stir and cook it well in the gravy.

Put the cheek and cowheel on a hot dish, and strain the gravy over them.

The bones can be used for soup.

Mock Hare.

- *Ingredients*—4 lb. shin of beef.
- 2 quarts of water.
- 2 carrots.
- 2 turnips.
- 1 onion.
- 6 cloves.
- 1 sprig of parsley, thyme, and marjoram.
- 1 glass of port wine.
- 3 oz. of flour.
- Pepper and salt.

Method.—Put the beef into the water with the vegetables cut in pieces, herbs, cloves, pepper and salt, and stew gently from four to five hours, until quite tender.

Then make a thickening of the flour, stir it in, and boil well for two or three minutes.

For serving, place the beef on a hot dish.

Add the wine to the gravy, and strain it over the meat.

Haricot Mutton.

- *Ingredients*—7 or 8 mutton cutlets.
- 1 pint of second stock.
- 1 carrot.
- 1 turnip.
- 1 43 onion.
- 1 stick of celery.
- 1 oz. of flour.
- Pepper and salt.
- 2 oz. of dripping.

Method.—Fry the cutlets a nice brown in the dripping.

Mix the flour smoothly with the stock; boil it in a stewpan for two minutes.

Then put in the cutlets and the vegetables cut in fancy shapes.

Stew gently for about three-quarters of an hour, until the meat and vegetables are tender.

Dish the cutlets in a circle; place the vegetables round them and pour the gravy over.

Sheep's Head.

- *Ingredients*—1 sheep's head.
- 1 oz. of butter or dripping.
- Pepper and salt.
- 1½ oz. of flour.
- A few drops of lemon juice.

Method.—See that the head has been properly prepared by the butcher, and the nostrils removed.

Soak it well in salt and water, and wash it carefully.

Cut out the tongue, remove the brains, and tie the head into shape with a piece of string.

Put it and the tongue into a saucepan of boiling water, and simmer it from three to four hours.

A quarter of an hour before it is cooked, put in the brains tied in muslin.

To make a sauce for it, melt the butter or dripping in a small stewpan.

Mix in the flour smoothly.

Pour in one pint of the broth from the sheep's head.

Stir and cook well, adding pepper and salt to taste a few drops of lemon juice, or one teaspoonful of vinegar.

44 Lastly, add the brains, chopped small.

For serving, put the head on a hot dish.

Remove the string, and pour the sauce over.

Sheep's Head au gratin.

- *Ingredients*—1 sheep's head.
- 2 tablespoonfuls of bread crumbs.
- ½ oz. of butter.
- 1 teaspoonful of chopped parsley.
- 1 teaspoonful of dried and powdered herbs.
- Lemon juice.
- Pepper and salt.

Method.—Boil the sheep's head according to the directions in preceding recipe.

When cooked, lay it on a greased baking-sheet.

Sprinkle over it the crumbs, parsley, and herbs, adding a few drops of lemon juice; pepper and salt.

Put the butter in little pieces about the head, and brown it in a quick oven or before the fire.

Serve with the brain sauce given in the foregoing recipe.

Liver and Bacon.

- *Ingredients* — 1 sheep's liver.
- 1 lb. of fat bacon.
- 1 pint of hot water.
- Some flour.
- Pepper and salt.

Method. — Cut the bacon into slices, and remove the rind.

Cut the liver into slices, and dip them in flour.

Fry the bacon in a frying-pan, then remove it, and fry the liver in the bacon fat, adding a little dripping, if necessary.

When the liver is cooked, place it on a hot dish; dredge the pan with about half an ounce of flour.

Fry the flour brown.

45 Then pour in one pint of boiling water, stir and boil for one or two minutes; adding pepper and salt to taste.

Place the liver in a circle in the middle of a hot dish.

Put the bacon round it, and strain the gravy over it.

Pigs' Fry, or Mock Goose.

- *Ingredients* — 1½ lb. of pigs' fry.

- 3 lb. of potatoes.
- 1 onion.
- 1 apple.
- A little sage.
- Pepper and salt.

Method.—Boil the potatoes until half-cooked.

Then cut them in slices.

Cut the fry in small pieces.

Chop the onion and apple small.

Dry and powder the sage leaves.

Grease a pie-dish, and put a layer of sliced potatoes at the bottom.

Place on them a layer of pigs' fry.

Sprinkle it with some of the onion, apple, and powdered sage, pepper, and salt.

Cover with another layer of potatoes; and put on that some more of the fry.

Sprinkle again with the onion, apple, pepper, and salt.

Proceed in this way until the dish is full, letting the last layer be potatoes.

Pour in half a pint of water; and cover the dish with a piece of pig's caul, or paper spread with dripping.

Bake in a moderate oven for one hour and a half.

It may be served in the pie-dish, or on a hot dish.

46

Mock Goose another way.

- *Ingredients*—1½ lb. of pigs' fry.
- Some dried and powdered sage.

- Chopped apple and onion.
- ¾ pint of cider.
- Pepper and salt.

Method. — Cut the fry in slices.

Thread the pieces on a long skewer.

Lay it on a greased baking-tin, and sprinkle with the onion, apple, sage, pepper, and salt, and cover with the caul.

Bake in a moderate oven until tender.

Then place the fry on a hot dish, and remove the skewer.

Make the cider boiling, and pour over the fry.

Tripe and Onions.

- *Ingredients* — 2 lb. of tripe.
- 3 good-sized onions.
- 2½ pints of milk.
- 2 oz. of flour.
- Pepper and salt.

Method. — Put the tripe into cold water, and bring it to the boil; this is to blanch it.

Blanch the onions likewise, then throw the water away, and cut the tripe into neat pieces.

Put them in the milk, with the onions cut in halves, and pepper and salt.

Stew gently for an hour.

Then take out the onions and chop them.

Remove the tripe, and put it on a hot dish.

Make a thickening of flour, and boil it well in the milk, and add the chopped onions.

47 Dish the tripe in a circle, and pour the milk and onions over.

Tripe may be cooked more economically by substituting water for milk.

Stewed Tripe.

- *Ingredients* — 2 lb. of tripe.
- 1 quart of brown sauce (*see* Sauces).

Method. — Blanch the tripe, as in the preceding recipe.

Simmer gently in brown sauce for two hours.

Dish in a circle, with the brown sauce poured over.

Broiled Steak.

Make the gridiron hot, and rub it with fat.

Lay the steak on it.

Place the gridiron close to a clear fire for about two minutes until the heat has scaled up that side of the steak.

Then turn it on to the other side, and let that remain close to the fire for the same length of time.

Then remove it further from the fire and cook more gradually, turning occasionally. It takes from ten to fifteen minutes to cook, according to the thickness of the steak.

Broiled Chop.

Cook like a steak. It will take from seven to ten minutes to cook. Serve very hot.

Fried Steak.

Make the frying-pan quite hot.

Put a little butter or fat in it, and make that quite hot also.

48 Put in the steak, and fry it over a quick fire for two minutes on one side, then turn it on to the other.

Moderate the heat applied, and cook gently for about twenty minutes, turning occasionally.

Savoury Roast.

- *Ingredients* — 1½ lb. of rump or beefsteak, cut thin.
- Some veal, or sage-and-onion, stuffing.
- ¾ oz. of flour.
- 1 cup of boiling water.

Method. — Lay the stuffing on the steak, roll it round it, and tie it with twine.

Place it in a pie-dish.

Pour the boiling water over it, and place another pie-dish, inverted, at the top of it.

Put it in a moderate oven for two or three hours, until the steak is tender.

Then put the steak on a hot dish.

Thicken the gravy with the flour and pour it over.

Breast of veal may be boned, and stuffed with veal stuffing and cooked in the same way.

Shoulder of Mutton Boned, Stuffed, and Rolled.

- *Ingredients* — 1 shoulder of mutton.

- Some veal stuffing, or sausage meat.

Method.—Remove the bone carefully, and place some stuffing in the place of it.

Roll up the mutton, and tie it firmly with twine.

It may be roasted, baked, or braised.

If braised, prepare it according to the directions given for braised breast of veal, using a large kettle, if a braising pan is not obtainable.

49

Braised Breast of Veal.

- *Ingredients*—3 or 4 lb. of breast of veal.
- Some veal stuffing.
- Some good second stock.
- Carrot, turnip, onion.
- Sprig of parsley, thyme, marjoram.
- 1 bay leaf.

Method.—Remove the bones from the veal, and put the stuffing in it.

Roll the veal round it, and sew it or tie it securely with twine.

Put the vegetables, cut in small pieces, in the bottom of a stew-pan.

Place the veal on them, and pour in sufficient stock to come half-way up it.

Put the lid on the stewpan, simmer gently until the veal is quite tender, allowing half an hour to each pound and half an hour over.

Then put the veal on a baking-sheet, and put in a quick oven to brown.

Strain the stock into a large stewpan, and boil it rapidly down to a glaze.

Put the veal on a hot dish, remove the string, and pour the glaze over it.

Place round the veal some carrot and turnip, cut in fancy shapes and cooked separately.

Toad-in-the-Hole.

- *Ingredients*—8 oz. of flour.
- 2 eggs.
- 1 pint of milk.
- 1½ lb. of ox kidney.
- A little salt.

Method.—Put the flour into a basin.

50 Make a well in the middle.

Put in the eggs; mix gradually.

Add the milk by degrees.

Beat well, and add the salt.

Cut the kidney in pieces, lay them in a well-greased Yorkshire-pudding tin; and pour the batter over.

Bake from one and a quarter to one and a half hours.

Irish Stew.

- *Ingredients*—2 lb. of potatoes.
- 1 lb. of scrag end of mutton.
- ½ lb. of onions.
- Pepper and salt.

Method.—Peel and slice the potatoes and onions, and cut the meat into small pieces.

Put a layer of meat in the bottom of a saucepan, then a layer of potatoes, then one of onions.

Season with pepper and salt, and continue placing the ingredients in the saucepan in alternate layers.

Pour in half a pint of water and stew gently, stirring occasionally, for about one hour and a half.

Sea Pie.

- *Ingredients*—2 lb. of steak.
- 2 onions.
- 1 carrot.
- 1 small turnip.
- ¾ lb. of flour.
- ¼ lb. of suet.
- 1 teaspoonful of baking powder.
- Pepper and salt to taste.
- Cold water.

Method.—Cut the vegetables and meat small, season them with pepper and salt, and put them into a large saucepan.

51 Put it by the side of the fire for the contents to simmer gently.

Chop the suet finely, add it to the flour and baking powder, and mix with cold water to a stiff paste. Roll it to the size of the saucepan.

Place it over the meat, and simmer gently for two hours.

For serving, remove the crust with a fish slice, put the meat and vegetables on to a hot dish, and place the crust on them.

Roast Bullock's Heart.

- *Ingredients*—1 bullock's heart.

- Some veal stuffing (double the quantity given in the recipe).

Method.—Wash the heart in salt and water, and cleanse it thoroughly.

Wipe it quite dry.

Cut off the flaps and fill the cavities with the stuffing.

Grease a piece of paper with dripping, and tie it firmly over the top of the heart to keep in the forcemeat.

Roast it according to the directions for roasting meat; it will take about two hours.

Gravy for the Heart.

- *Ingredients*—1 pint of stock.
- The trimmings from the heart.
- 1 onion.
- 1 oz. of butter.
- 1 oz. of flour.
- A little Harvey's sauce or catsup.
- A little burnt sugar, if necessary, for colouring.

Method.—Put the trimmings into a saucepan with the onion and water, and simmer gently while the heart is cooking.

Then melt the butter in a stewpan.

52 Mix in the flour smoothly; add the liquor strained.

Stir and boil three minutes; add the sauce, pepper and salt, and colouring.

Put the heart on a hot dish, remove the paper, and pour the gravy round it.

If preferred, the heart may be baked.

53

SAUCES.

Sauces are often failures, chiefly because they are not made of a proper consistency; and because the flour in them is not sufficiently cooked. It should be remembered that the starch in flour wants to be *well boiled*, otherwise it will be indigestible, and the sauce will have a raw, pasty taste. A sauce is not ready when it *thickens*, but should be boiled for quite three minutes. Its consistency should depend on what it is to be used for. Ordinary sauces, served in a sauce tureen, should be fairly thick; the proportions taken should be 1 oz. of butter; ¾ oz. of flour; ½ pint of milk. If the sauce is to be used to coat anything very thinly (new potatoes, for example), ½ oz. of flour, instead of ¾ oz., would be sufficient. If a sauce is required to entirely mask a small piece of fish, or chicken, &c., 1 oz. of flour should be used, with the proportions of milk and butter already given. Every ingredient should be properly weighed or measured. Carelessness in this respect is a mark of ignorance, and *must* occasion failures.

For making most of the ordinary sauces, the butter is melted first in a small stewpan, care being taken that it does not discolour; the flour is then mixed with it. If the mixing is not perfect, the sauce will be lumpy. The milk, stock, or water, is then poured in, and the sauce is stirred *one way*, until it has boiled three minutes. If cream is used, it is then added, and allowed just to boil in the sauce.

In making economical sauces, when less butter and flour are used (*see* Economical Family Sauce), the method employed is different. The flour is then mixed very smoothly 54 with a little of the milk, water, or whatever is used, and then added to the remainder, which may be cold or boiling; but greater care is required to keep it smooth when the liquid is poured in boiling.

English Melted Butter.

- *Ingredients* — 1 oz. of butter.
- ¾ oz. of flour.
- ½ pint of water.

- Pepper and salt.

Method. — Melt the butter in a small stewpan.

Mix in the flour smoothly.

Add the water; stir and cook well.

Then add pepper and salt, and it is ready to serve.

Plain White Sauce.

- *Ingredients* — 1 oz. of butter.
- ¾ oz. of flour.
- ½ pint of milk.
- A few drops of lemon juice.
- Pepper and salt.

Method. — Melt the butter in a small stewpan.

Mix in the flour smoothly.

Add the milk.

Stir and cook well.

Then add the lemon juice and seasoning.

A little cream may also be added if desired.

Maître d'Hôtel Sauce.

- *Ingredients* — ¾ oz. of butter.
- ½ oz. of flour.
- ½ pint of milk.
- A few drops of lemon juice.
- Pepper and salt.
- A teaspoonful of finely-chopped parsley.

55 *Method.* — Melt the butter in a small stewpan.

Mix in the flour smoothly.

Add the milk; stir and cook well.

Then add the lemon juice, seasoning, and chopped parsley.

Mayonnaise Sauce.

- *Ingredients* — 2 yolks of eggs.
- 1 gill of salad oil.
- 2 tablespoonfuls of taragon vinegar.
- Pepper and salt.

Method. — Put the yolks, which must be perfectly free from the whites, into a basin, which in summer time should be placed on ice.

Work them well with a whisk or wooden spoon, adding the oil drop by drop.

When the sauce is so thick that the whisk, or spoon, is moved with difficulty, the oil may be added more quickly, but still very gradually.

Lastly, add the taragon vinegar and seasoning.

Note. — Success in making this sauce depends on first dividing the yolks completely from the whites. Secondly, in keeping them and the oil quite cold. Thirdly, on adding the oil, drop by drop, until the sauce is perfectly thick. If the sauce is made in a warm place, or the oil mixed in too quickly, it is apt to curdle. Should this occur, put a yolk in another basin and very slowly add the sauce to it, stirring briskly; this will generally make it smooth again. Two yolks will be sufficient for any quantity of sauce, taragon vinegar being added in proportion to the oil used.

Tartare Sauce.

- *Ingredients*—2 yolks.
- ¼ pint of salad oil.
- 2 tablespoonfuls of taragon vinegar.
- 1 teaspoonful of finely-chopped parsley.
- A 56 few capers, or a chopped gherkin.
- Pepper and salt.
- If liked, a teaspoonful of ready-made mustard.

Method. —Proceed as in making Mayonnaise Sauce; adding when the sauce is ready the parsley, capers, mustard, and seasoning.

Egg Sauce.

- *Ingredients*—1 oz. of butter.
- ¾ oz. of flour.
- ½ pint of milk.
- Lemon juice.
- Pepper and salt.
- 1 or 2 hard-boiled eggs.

Method.—Melt the butter in a small stewpan.

Mix in the flour smoothly.

Add the milk, and stir and cook well.

Then add the lemon juice, seasoning, and the chopped whites of the eggs.

If a very thick sauce is required, take 1 oz. of flour. Cream may be added if desired.

Brown Sauce.

- *Ingredients*—2 oz. of butter.
- 1½ oz. of flour.
- A small piece of carrot, turnip, and onion.
- A few button mushrooms.
- 1 pint of good stock.
- A few drops of lemon juice.
- Seasoning to taste.

Method.—Put the butter into a stewpan and fry the vegetables in it.

Then mix in the flour and fry that.

Add the stock; stir and cook well.

57 Squeeze in the lemon juice, and add the seasoning.

Strain through a tammy-cloth or fine strainer.

Genoise Sauce.

- *Ingredients*—1 oz. of butter.
- ¾ oz. of flour.
- 1½ gills of stock.
- ½ wineglass of port.
- A tiny piece of carrot, turnip, and onion.
- ½ teaspoonful of anchovy sauce.
- ½ teaspoonful of Harvey's sauce.
- Pepper and salt.

Method.—Melt the butter in a small stewpan, and fry the vegetables in it.

Then add the flour, and fry that.

Pour in the stock; stir and cook well.

Then add the wine and other ingredients,

Stir until it boils again, and then strain it.

Béchamel Sauce.

- *Ingredients*—2 oz. of butter.
- 1½ oz. of flour.
- 1 pint of good white stock.
- ¼ pint of cream.
- A few drops of lemon juice.
- Pepper and salt.

Method.—Melt the butter in a stewpan.

Mix in the flour smoothly.

Add the stock.

Stir and cook well.

Then stir in the cream; let it boil in the sauce; and add lemon juice, pepper, and salt.

Strain through a tammy-cloth.

58 Milk may be substituted for the white stock, if more convenient. To flavour it, a small piece of carrot, turnip, and onion, and 6 button mushrooms should be boiled in it.

Sauce Hollandaise.

- *Ingredients*—¼ pint of plain white sauce.
- The yolks of 4 eggs.
- A little cayenne pepper and salt.
- A few drops of lemon juice, or taragon vinegar.

Method.—Put the white sauce and eggs into a jug, which must be placed in a saucepan of boiling water.

Stir until the mixture thickens, being careful it does not curdle.

When quite ready, add the lemon juice or vinegar.

Lobster Sauce.

- *Ingredients*—1 small lobster.
- Some spawn.
- 1½ oz. of butter.
- 1 oz. of flour.
- ½ pint of milk.
- ½ gill of cream.
- A few drops of lemon juice.
- Pepper and salt.

Method.—Remove the flesh from the body and claws of the lobster, and cut it in small pieces.

Then boil the shell, broken small, in the milk.

Rub the spawn with ¼ oz. butter through a hair sieve.

Melt the remaining butter in a small stewpan.

Mix in the flour smoothly, and then add the milk, strained.

Stir until it thickens.

Put in the spawn and butter, and continue stirring until the flour is well cooked.

59 Then add the cream—let it boil in the sauce—and lastly, the lemon juice, pepper and salt, and lobster.

Lobster Sauce (a plainer Receipt).

- *Ingredients*—Part of a tin of lobster.

- 1 oz. of butter.
- 1 oz. of flour.
- ¾ pint of milk.
- A few drops of lemon juice, or ½ a teaspoonful of vinegar.
- Pepper and salt.

Method.—Cut up the lobster.

Melt the butter in a small stewpan.

Mix in the flour smoothly.

Add the milk; stir and cook well.

Then add the lemon juice, seasoning, and pieces of lobster.

Shrimp Sauce.

Remove the heads, tails, and skin from half a pint of shrimps; prepare some sauce as directed in the first or second recipe for lobster sauce, substituting the shrimps for the lobster.

Oyster Sauce.

- *Ingredients*—1 oz. of butter.
- 1 oz. of flour.
- ½ pint of milk.
- 1 dozen of oysters.
- ½ gill of cream.
- A few drops of lemon juice.
- Salt, pepper, and a little cayenne.

Method.—Remove the beard and white part of the oysters, and cut each one in two.

Strain the liquor through muslin, and scald the oysters in it (*i.e.* put the liquor, with the oysters in it, in a saucepan, and just bring it to the boil).

60 Put the beards and hard white parts in the milk and simmer them to extract the flavour.

Then melt the butter in a small stewpan.

Mix in the flour smoothly.

Strain in the milk and oyster liquor, and stir and cook well.

Then add cream, and stir until the sauce again boils.

Lastly, add the oysters, pepper, salt, and lemon juice.

French Sauce.

- *Ingredients*—1 oz. of butter.
- ½ oz. of flour.
- 1 gill of milk.
- 1 gill of cream.
- The yolk of one egg.
- Pepper and salt.

Method.—Melt the butter in a small stewpan.

Mix the flour smoothly.

Add the milk, stir and cook well.

Pour in the cream and let it boil in the sauce. Then take it off the fire, and mix in the yolk of the egg.

Add pepper and salt to taste.

Celery Sauce.

- *Ingredients*—1 oz. of butter.
- 1 oz. of flour.
- 2 tablespoonfuls of cream.
- ¾ pint of white stock or milk.
- 1 head of celery.

Method. — Boil one head of celery in ¾ of a pint of white stock or milk.

When tender, strain it from the liquor and rub it through a hair sieve.

Melt the butter in a small stewpan.

Mix in the flour smoothly.

61 Add the stock or milk; stir and cook well.

Pour in the cream, and stir until the sauce boils again.

Add pepper and salt to taste.

Tomato Sauce.

- *Ingredients* — 6 ripe tomatoes.
- ¼ lb. of bacon.
- 1 oz. of flour.
- A piece of carrot, turnip, and onion.
- A sprig of parsley.
- Thyme, marjoram, and a bay leaf.
- A teaspoonful of vinegar.
- Pepper and salt.

Method. — Cut the bacon in slices and fry it.

Then put in the vegetables and fry them, dredge in the flour, and then add the tomatoes and fry them lightly.

Empty the contents of the frying-pan on a hair sieve, and rub the tomatoes through. The hair sieve will keep back the other vegetable, the flavour of which only is wanted.

Add the vinegar and seasoning, and make the sauce hot.

Onion Sauce.

- *Ingredients*—4 or 5 fair-sized onions.
- ½ pint of plain white sauce or melted butter (1st recipe).

Method.—First, blanch the onions by putting them in cold water and bringing it to the boil.

Throw the water away.

Put the onions in fresh water and boil for an hour, or an hour and a half, until tender.

Chop them finely and add them to the sauce or melted butter.

62

Soubise Sauce.

- *Ingredients*—½ pint of plain white sauce.
- 2 tablespoonfuls of cream.
- 4 or 5 onions.

Method.—Blanch the onions (as in preceding recipe) and boil until tender.

Then rub through a hair sieve.

Make some plain white sauce (*see* recipe), and add to it the cream and pulped onion.

Bread Sauce.

- *Ingredients*—2 oz. of bread crumbs.
- ½ pint of milk.
- 6 peppercorns.
- 2 tablespoonfuls of cream, or ½ oz. of butter.

- A small piece of onion.

Method.—Steep the onion and peppercorns in the milk, and put the milk on to boil.

Then remove the onions and peppercorns, and sprinkle in the crumbs.

Set the sauce by the side of the fire for six minutes, and then heat to boiling point, adding either the cream or butter.

Salt must be added to taste; also a little cayenne.

Economical Family Sauce.

- *Ingredients*—¼ lb. of flour.
- 1 pint of milk.
- 1 pint of water.
- 1½ oz. butter.

Method.—Mix the flour very smoothly with a little water.

Put the rest of the water, with the milk and butter, in a saucepan on the fire to boil.

63 When it boils, put in the flour, stirring until the sauce is cooked.

Add pepper and salt to taste. If liked, a few drops of lemon juice or vinegar may be added.

This sauce will form the basis of many other plain sauces: To use with fish, put in a tablespoonful of anchovy. Onion sauce is made by adding cooked and chopped onions when the sauce is ready. Caper sauce, by adding capers; or, as a substitute, chopped gherkin.

This sauce may be made still more economically by using water only instead of milk.

Wine Sauce.

- *Ingredients*—1 oz. of lump sugar.
- ¼ pint of water.
- A wineglass of sherry.
- A few drops of cochineal.
- A dessertspoonful of jam.

Method.—Boil the sugar and water together until reduced to one half.

Add the jam; let it melt.

Then add the sherry and cochineal, and strain.

Piquant Sauce.

- *Ingredients*—½ pint of brown sauce.
- 1 tablespoonful of capers.
- 1 tablespoonful of chopped gherkin.
- 1 tablespoonful of very finely chopped shalot.
- ¼ pint of vinegar.
- Pepper and salt.

Method.—Simmer the shalot, capers, and gherkin, in the vinegar until the shalot is quite soft.

Pour in the sauce, and let it boil up.

Season to taste.

64

Sauce Réforme.

- *Ingredients*—1 pint of brown sauce.

- 1 wineglass of port wine.
- 1 teaspoonful of anchovy sauce.
- 1 teaspoonful of Harvey's sauce.
- 2 tablespoonfuls of red-currant jelly.

Method.—Boil all the ingredients together, and the sauce is ready.

Port-wine Sauce for Wild Duck.

- *Ingredients*—2 wineglasses of port.
- Juice of half a lemon.
- 1 finely chopped shalot.

Method.—Boil altogether and strain.

Sweet Sauce.

- *Ingredients*—1 teaspoonful of arrowroot.
- Juice of half a lemon and a little rind.
- 2 tablespoonfuls of castor sugar.
- ½ pint of water.

Method.—Put the water with the lemon-rind and sugar into a saucepan to boil.

Mix the arrowroot smoothly with a little cold water.

When the water in the saucepan boils, pour it in and stir it until it thickens; then strain it and add the lemon juice.

A glass of sherry may be added to this sauce if desired.

German Sauce.

- *Ingredients* — The yolks of 2 eggs.
- 1 wineglass of sherry.
- 1 dessertspoonful of castor sugar.

65 *Method.* — Put all the ingredients into a saucepan, and mill over the fire with a whisk until the sauce froths.

For a *Christmas Pudding* make the sauce with three yolks, and a wineglass of brandy.

A Nice Sweet Sauce.

- *Ingredients* — ½ pint of plain white sauce or melted butter (omitting the seasoning).
- 1 wineglass of sherry or brandy.
- 2 dessertspoonfuls of castor sugar.

Method. — Add the wine and sugar to the sauce, and it is ready for use.

Jam Sauce.

- *Ingredients* — 3 tablespoonfuls of red jam.
- ½ pint of water.
- 1 oz. of lump sugar.
- Juice of half a lemon.

Method. — Boil the jam, sugar, and water together for three minutes.

Add the lemon juice, and strain.

The lemon may be omitted if the flavour is not liked.

Apple Sauce, No. 1.

- *Ingredients* — 6 good-sized apples.
- 1 oz. of butter.
- 1 tablespoonful of moist sugar, or more, according to taste.
- ½ gill of water.

Method. — Wash the apples and slice them, but do not peel or core them.

Put them in a stewpan with the water, butter, and sugar.

Stew gently for about thirty minutes, stirring occasionally.

66 Rub them quickly through a hair sieve, and put the sauce in a hot tureen.

The hair sieve keeps back the rind and pips.

Apple Sauce, No. 2.

- *Ingredients* — 6 large apples.
- 1 oz. of butter.
- 1 tablespoonful or more of moist sugar.
- ½ gill of water.

Method. — Peel, core, and slice the apples.

Stew them with the water, sugar, and butter until tender.

Then beat to a pulp with a fork.

Mint Sauce.

- *Ingredients* — 3 tablespoonfuls of finely-chopped fresh mint.

- 1 tablespoonful of sugar.
- ¼ pint of vinegar.

Method.—Mix all together, and let the sauce stand for an hour before serving.

Horse-radish Sauce.

- *Ingredients*—1 stick of horse-radish.
- ½ gill of cream.
- 1 tablespoonful of vinegar.
- ½ gill of milk.
- 1 teaspoonful of ready-made mustard.
- 1 teaspoonful of castor sugar.
- Pepper and salt.

Method.—Scrape the horse-radish finely, and mix with all the other ingredients.

If cream is not to be had, use milk thickened with a little corn-flour. But it is not so good.

67

Gravy for Made Dishes.

- *Ingredients*—1 lb. of gravy beef.
- 1 quart of water.
- A piece of onion, carrot, and turnip.
- 1 sprig of parsley.
- Thyme and marjoram.
- Pepper and salt to taste.

Method.—Cut the beef into small pieces.

Put it with the vegetables into a stewpan with the water, and simmer very gently for four hours; then strain.

If a thick gravy is required, thicken with one and a half ounces of flour; add pepper and salt to taste.

To this gravy may be added a little sauce, catsup, port or sherry wine, &c., according to the purpose for which it is required.

Scraps of cooked meat and bones may be substituted for the fresh meat where economy must be studied.

Glaze.

Boil down one or two quarts of second stock (which will jelly when cold) until it is quite thick, and coats a spoon. One quart may be boiled down to a quarter of a pint.

Pour it into a jar.

When wanted for use, put the jar to stand in a saucepan of boiling water until it is dissolved.

Glaze is used for enriching gravies and soups, and for glazing meat.

Cheap Glaze for Meat.

- *Ingredients*—3 teaspoonfuls of Liebig's Extract of Meat.
- ½ oz. of Nelson's or Swinborne's Gelatine, or isinglass.
- Pepper and salt.
- ½ pint of cold water.

68 *Method.*—Soak the gelatine in the water for three-quarters of an hour.

Add the meat extract, and pepper and salt.

Stir and boil until reduced to about a quarter of a pint.

This glaze can only be used for glazing meat.

Béarnaise Sauce.

- *Ingredients*—1 finely-chopped shalot.
- ½ gill of white sauce.
- 1 tablespoonful of taragon vinegar.
- The yolks of 4 eggs.
- 1 dessertspoonful of finely-chopped parsley.
- Pepper and salt.

Method.—Put the shalot and vinegar into a saucepan; boil until the vinegar has evaporated, but do not let the shalot burn.

Add the eggs and sauce, and mill with a whisk until the eggs are thick.

Add the parsley and pepper and salt.

69

BREAKFAST DISHES AND BEVERAGES.

Oatmeal Porridge.

- *Ingredients* — ½ lb. of coarse oatmeal.
- 1 quart of water.

Method. — Put the water on to boil.

When boiling, sprinkle in the oatmeal, stirring all the time.

When it thickens, put it by the side of the fire, and stir occasionally.

Cook it for quite three-quarters of an hour, longer if possible.

When the time can be allowed, three hours will not be too long a time, especially if the porridge is for anyone with a weak digestion.

A better plan is to put the saucepan containing it, after the contents have boiled for ten minutes, to stand in a saucepan of briskly boiling water; it will then cook without danger of burning, and may be left for any length of time; care only being taken that the water in the under saucepan does not boil away.

Whole-meal Porridge.

This may be made in the same way as oatmeal, but it requires even longer cooking.

Dry Toast.

Cut the bread into rather thin slices, and remove the crust.

70 Toast it slowly, holding it at a little distance from a bright clear fire.

When ready, put it at once into the rack; because, if the toast is placed flat on a table, it loses its crispness.

The crusts may be soaked for plain puddings, or dried and powdered for bread crumbs.

Buttered Toast.

Cut the bread about half an inch in thickness.

Toast quickly in front of a clear fire.

Put the butter on directly the toast is taken off the fork, and spread it quickly.

Put the toast on a *hot* plate, and take care that it is served hot.

Toasted Bacon.

Cut the bacon in thin slices, and toast it in a small Dutch oven or on a toasting fork until the fat is transparent.

Fried Bacon.

Cut the bacon in thin slices, and fry it in its own fat. It will be cooked when the fat is transparent. It must not be cooked too quickly, or the fat will burn up and be wasted.

Eggs and Bacon.

Toast or fry the bacon, and lay a nicely poached egg on each slice.

Boiled Eggs.

Put the eggs into boiling water, and boil an ordinary sized egg for three minutes; new-laid eggs will take one minute longer. Eggs boiled five minutes will be nearly hard. To make them quite firm, boil them steadily for ten minutes. To make them mealy, boil them for an hour.

71

Poached Eggs.

Eggs for poaching should be perfectly *fresh*, or they will not keep a nice shape.

Let the water be quite boiling; add to it a little salt.

Break the eggs into cups, and slip them gently into the boiling water.

As soon as the white is nicely set, remove them with a fish slice.

Trim the eggs neatly, and serve them on hot buttered toast.

An egg-poacher will be found very convenient for cooking eggs this way.

Fried Kidneys.

- *Ingredients*—A few kidneys.
- A little butter or dripping.
- A little flour.
- Some gravy.
- Pepper and salt.

Method.—Split open the kidneys lengthwise.

Flour them and fry them slowly in the butter or dripping for about four minutes.

Dish them on pieces of toast.

Pour the gravy into the pan; stir and boil for a minute, and then strain round the kidneys.

Kidneys Toasted.

- *Ingredients*—Some kidneys.
- Toasted bread.

Method.—Split open the kidneys lengthwise.

Toast them before a clear fire; when the gravy ceases to drop red they will be sufficiently cooked.

A *hot* dish should be placed under them to catch the gravy.

72 Place the toast on the dish and put the kidneys on it, and sprinkle over them a little pepper and salt.

Stewed Kidneys.

- *Ingredients*—2 or 3 kidneys.
- ½ pint of nice gravy.
- 1 dessertspoonful of flour.
- Pepper and salt to taste.
- Lemon juice.

Method.—Mix the flour smoothly with the gravy.

Put it into a stewpan, and boil well for three minutes.

Put in the kidneys cut in slices, and simmer gently for about fifteen minutes.

Add a squeeze of lemon juice; pepper and salt to taste.

Serve on a piece of toast, and pour the gravy over.

Stuffed Kidneys.

- *Ingredients* — 3 or 4 kidneys.
- ½ oz. of butter.
- Half a shalot, chopped finely.
- 1 dessertspoonful of parsley.
- 1 tablespoonful of bread crumbs.
- A few drops of lemon juice.
- A little cayenne.
- Pepper and salt.

Method. — Toast or broil the kidneys and split them open.

Fry the shalot in the butter.

Mix in the bread crumbs and parsley; add lemon juice, cayenne pepper, and salt.

Lay a little of the stuffing in each kidney and fold it over.

Serve very hot.

73

Kidneys à la Tartare.

- *Ingredients* — A few kidneys.
- ½ pint of Tartare sauce.

Method. — Split the kidneys open, and toast or broil them nicely.

Serve on toasted bread with Tartare sauce in a tureen.

Fried Sausages.

- *Ingredients* — Sausages.
- A little butter or dripping.

- Some toasted bread.

Method.—Prick the sausages with a fork, and fry them with butter or dripping, turning them that they get browned equally.

Serve them on toasted bread, with some nice gravy in a sauceboat.

Some people like the toast soaked in the fat in the pan, but this is a matter of taste.

Baked Sausages.

Prick the sausages, and place them on a greased baking-sheet.

Bake until they are nicely browned.

Serve on toast, with gravy in a sauceboat.

If liked, the toast can be soaked in the fat that runs from the sausages.

Oxford Sausages.

Remove the sausage-meat from the skins, and place it in little rough heaps on a greased baking-sheet.

Bake in a quick oven until browned.

Serve on toast.

74

Tomatoes stuffed with Sausage Meat.

- *Ingredients*—Some nice ripe tomatoes.
- Some sausage meat.

Method.—Cut the stalks from the tomatoes, but do not take out any of the inside.

Heap a little sausage meat on the top of each tomato.

Put them on a greased baking-sheet, and bake in a moderate oven for about fifteen minutes.

Croustards with Minced Meat.

- *Ingredients*—Some stale bread.
- Scraps of cold meat.
- A little nice gravy.
- A little mushroom catsup.
- Pepper and salt to taste.

Method.—Cut the bread into slices three-quarters of an inch in thickness.

Stamp it into rounds with a circular cutter.

Mark the middle with a cutter two sizes smaller, and scoop out the inside, making little nests of them, and taking care not to break the bottom or sides.

Fry the cases in hot fat (*see* French Frying); drain them and put them inside the oven to keep hot.

Mince the meat nicely, removing skin and gristle.

Make a little gravy hot in a stewpan.

Put in the mince, and make it hot without letting it boil.

Flavour to taste with catsup, pepper and salt.

Fill the croustard cases and serve immediately: they should be placed on a folded napkin, and garnished with parsley.

Mince à la Reine.

- *Ingredients*—1 dozen mushrooms.
- Some slices of cold meat.
- (Cold 75 game or chicken are excellent for this purpose).
- 6 eggs.
- Some rounds of bread, toasted or fried.
- 1 pint of good gravy.
- Pepper and salt to taste.

Method.—Peel the mushrooms.

Wash and dry them well, and cut them in slices.

Put them in a stewpan with part of the gravy, to stew for about thirty minutes, until they are tender.

Mince the meat and make it hot in a saucepan, with enough gravy to moisten it, adding pepper and salt to taste.

Poach the eggs nicely, and fry or toast the bread (fried bread is best).

Put the slices of fried bread on a hot dish; cover each piece with the minced meat, and lay an egg on each.

Pour the gravy and mushrooms round, and serve very hot.

As a decoration, a tiny pinch of finely-chopped parsley might be put on the top of each egg.

Sheep's Head Moulded.

- *Ingredients*—1 sheep's head.
- 2 hard-boiled eggs.
- Pepper and salt.

Method.—Clean, and then boil the head until the flesh will leave the bones easily.

Take out all the bones; cut the meat into pieces an inch in size, and season them well with pepper and salt.

Cut the eggs into slices, and place them round the top of a cake-tin or basin.

Put in the head, and put a weight on it to press it down.

When cold turn it out; serve garnished with parsley.

76

Veal Cake.

- *Ingredients* — Remains of cooked veal.
- Slices of ham.
- 2 or 3 hard-boiled eggs.
- Some nice second stock.
- A little gelatine.
- Some forcemeat balls.

Method. — Butter well a plain mould or basin.

Decorate it with slices of egg, and balls made of veal forcemeat.

Cut the ham and the veal into neat pieces.

Season them well with pepper and salt, and, if liked, a little chopped parsley.

Place them in the mould, and fill it up with stiff second stock.

If the stock is not stiff enough, mix with it a little melted gelatine.

Cover the mould, and bake for one hour in a moderate oven.

Let it get cold, and then turn it on to a dish.

Brawn.

- *Ingredients* — 1 pig's head.

- 2 or 3 hard-boiled eggs.
- 2 onions.
- 6 cloves.
- 1 blade of mace.
- 2 dozen peppercorns.
- 1 sprig of parsley, thyme, and marjoram.

Method.—Clean the head well, and pickle it for three days (*see* Pickle for Meat).

Then put it in enough cold water to cover it, and boil it gently for three hours or more, until the flesh will leave the bones easily.

Take out the tongue, skin it, and cut it in slices.

Stamp them into fancy shapes with a paste cutter; wet 77 a plain round mould and decorate it with them and the eggs cut in slices.

Remove the meat from the bone, and cut it into large dice.

Take one quart of the liquor in which the head was boiled; put the bones into it, with the peppercorns, cloves, onions, and herbs; boil down for half an hour with the lid off the saucepan.

Then strain one pint of the broth into another saucepan.

Season the pieces of meat with pepper, and a little salt if necessary; put them into the broth.

Let it come to the boil, and then pour it into the decorated mould.

When set, turn it on to a dish.

Scalloped Eggs.

- *Ingredients*—Some eggs.
- Bread-crumbs.
- A little onion, chopped as finely as possible (this may be omitted, if liked).
- A little finely-chopped parsley.
- Pepper and salt to taste.

Method.—Grease some deep scallop shells.

Dust them over with bread crumbs, mixed with the parsley and onion.

Put an egg into each shell, and sprinkle with more crumbs, parsley, onion, pepper and salt.

Put them into a brisk oven until set.

Eggs sur le Plat.

- *Ingredients*—4 eggs.
- ½ oz. of butter.
- Pepper and salt.

Method.—Take a dish that will stand the heat of the oven; melt the butter in it.

Break the eggs on to it very carefully.

78 Pepper and salt them, and put them into the oven until they are set.

They must be served on the same dish.

Buttered Eggs.

- *Ingredients*—1 piece of fried or toasted bread.
- 1 tablespoonful of gravy.
- 1 oz. of butter.
- Pepper and salt.
- 4 eggs.

Method.—Break the eggs into a basin, and add to them the gravy, pepper, and salt.

Melt the butter in a small frying or omelet pan; pour in the eggs, and stir quickly up from the bottom of the pan, until the whole is a soft yellow mass.

Spread on the toast, and serve very quickly.

Egg Croustards.

- *Ingredients*—Some slices of stale bread, about ¾ inch in thickness.
- Some eggs.
- Some nicely-flavoured gravy.

Method.—Stamp out some rounds of bread with a circular paste-cutter.

Mark the middle with one a size smaller.

Then with a knife scoop out the inside, making little nests of bread, taking care not to break the bottom or sides.

Fry these cases in hot fat (*see* French Frying).

When fried, drain them on kitchen paper, and keep them hot.

Make some water boiling hot in a stewpan; add to it a little lemon juice.

Put into it the eggs broken gently into cups.

Poach until the whites are set, then remove them carefully with a fish slice, and put an egg into each croustard.

Place them on a hot dish, and pour gravy boiling hot over them.

79

Eggs and Anchovy.

- *Ingredients*—2 eggs.
- 1 slice of fried or toasted bread.

- A little anchovy paste.
- 1 oz. of butter.
- Pepper and salt to taste.

Method. — Let the fried or toasted bread be quite hot (fried bread is the best), and spread it thinly with anchovy paste.

Make the butter quite hot in a frying or omelet pan.

Break the eggs into it, add pepper and salt, and stir very quickly, until they are a soft yellow mass.

Spread it quickly over the toast, and serve immediately.

Eggs in Cases.

- *Ingredients* — 4 tablespoonfuls of bread crumbs.
- 1 dessertspoonful of finely-chopped parsley.
- Pepper and salt.
- If liked, a boiled onion very finely chopped.
- 6 eggs.
- 6 paper cases.

Method. — Butter well some paper cases; mix the crumbs, parsley, onion, pepper, and salt together; put a little at the bottom of each case.

Break the eggs gently, and put one egg into each case.

Cover each with some of the crumbs and seasoning, and put the cases in a quick oven to bake until the eggs are set.

Broiled Mushrooms.

Choose nice large mushrooms; peel and wash them, and wipe them dry.

Cut out the stems, and put them, with the top of the mushrooms downwards, on a gridiron.

80 Put a small piece of butter on each, and broil for ten minutes slowly.

Remove them carefully, as the mushrooms will be by that time full of delicious gravy.

Broiled Dried Haddock.

Soak it in cold water for an hour before using.

Broil it slowly over a clear fire until it is quite hot, turning occasionally.

Rub some butter over it, and serve it at once.

Bloaters.

Cut the bloaters open down the back, and bone them.

Lay them one on the other with the insides together.

Broil them slowly over a clear fire, turning occasionally.

Serve very hot, with a little butter rubbed over them.

If preferred, they may be broiled unboned.

Red Herrings.

Remove their heads and tails.

Slit them open down the back and remove the bone.

Egg and bread-crumb them, and broil them over a clear fire.

If preferred, they may be broiled unboned.

Tea.

Measure a teaspoonful of tea for each person, and one teaspoonful over.

Make the teapot quite hot by filling it with boiling water; let it stand in it for three minutes; then empty the teapot.

Put in the tea, and pour boiling water over it.

Cover it with a tea-cosy, and let it infuse for five minutes before using. The longer it stands, the darker it 81 will get; but for people of weak digestions, it should be used after five minutes' infusion only.

The water should be fresh spring water, and should be used as soon as it boils. Water that has been boiled for any length of time is flat from the loss of its gases.

Coffee.

To have coffee to perfection it should be freshly roasted and ground, as coffee quickly loses its flavour. If this is not possible, use the best French coffee sold in tins. The water should be freshly boiled; the coffee itself should *not* be *boiled*, but only infused in the boiling water. Boiling disperses the aroma. It can, however, be made more economically if boiled, and therefore recipes are given for its preparation in this manner. Chicory is generally used with coffee in the proportion of two ounces of chicory to one pound of coffee.

Coffee (Soyer's method.)

- *Ingredients*—3 oz. of coffee.
- 1 pint of boiling water.

Method.—Put the coffee into a clean stewpan.

Stir over the fire until it smokes, but do not let it burn.

Then pour in the boiling water.

Cover close, and set by the side of the fire for ten minutes.

Strain through thick muslin.

Coffee (another method).

- *Ingredients* — 3 oz. of coffee.
- 1½ pint of boiling water.

Method. — Make a jug hot.

Put the coffee in it, and pour over the boiling water.

Let it stand in a hot place for half an hour.

Then strain through thick muslin.

82

Café au Lait.

Half fill a cup with nicely-made coffee, and pour in the same quantity of boiled milk.

Coffee (economical method).

- *Ingredients* — ¾ lb. of coffee.
- 2 quarts of cold water.

Method. — Make a bag of rather thick muslin, and put the coffee into it. The bag should be rather large, so that the coffee will have plenty of room.

Tie the ends of the bag securely.

Put it into a saucepan with the water; bring to the boil, and boil steadily for one hour.

Strain through thick muslin.

This will make strong coffee, which can be diluted with boiling water as required.

Coffee made in a Percolator.

- *Ingredients*—3 oz. of coffee.
- 1½ pint of boiling water.

Method.—Make the percolator hot.

Put the coffee in it, and pour on the boiling water.

Let it stand in a hot place for about ten minutes.

Cocoa.

This is best, especially for invalids, if prepared from the nibs; these should be perfectly fresh.

Put a quarter of a pound of nibs into two quarts of cold water; simmer for five hours and then strain.

When cold remove the fat; heat it as required.

Cocoa may also be made from any of the different preparations.

Make it according to directions given on the canisters 83 , and be very careful to mix it thoroughly. Nothing is so unpleasant as to have the sides and bottom of the cup coated with cocoa.

It is better to prepare it in a small saucepan; it should be boiled for two or three minutes.

It is more nourishing if mixed with milk instead of water.

Chocolate.

This is only a thicker preparation of cocoa, and may be made in the same way.

84

COLD MEAT COOKERY.

Hash.

- *Ingredients*—The remains of cold meat.
- Some nice stock or gravy.
- Flour, in the proportion of ½ oz. to every ½ pint of gravy.
- Pepper and salt, and, if liked, a little catsup, or Harvey's sauce.
- Toasted or fried bread.

Method.—Cut the meat into neat pieces.

Mix the flour smoothly with the gravy, and boil for three minutes, stirring all the time.

Add seasoning and catsup or a little sauce.

Then put the pieces of meat into the gravy and let them warm through; but do not let the gravy *boil* when the meat is in it, as that would toughen it.

Tinned oysters make a nice addition to a hash.

For serving, put the hash on a hot dish and garnish with sippets of fried or toasted bread.

If no gravy or stock is available, make some by breaking up any bones from the meat; boil them in a sufficient quantity of water, with a piece of carrot, turnip, onion, celery, and a small bunch of herbs.

Boil for quite an hour, and then strain the liquor.

Minced Meat.

- *Ingredients*—Some scraps of cold meat.

- A little gravy.
- Some boiled rice or potatoes.
- Pepper and salt to taste.

85 *Method.* — Mince the meat finely with a knife, or mincing machine (the flavour is nicer if a knife is used).

Mix with sufficient gravy to moisten the meat, and stir over the fire until hot; but do not let the gravy boil.

Serve with a border of boiled rice, or mashed potatoes round it.

If veal or chicken is minced, squeeze in a few drops of lemon juice, and serve with sliced lemon.

A little cooked ham should be added to these minces, to give them flavour; minced beef is improved by the addition of a few oysters.

Mince (with Eggs).

Prepare some mince, as in preceding recipe, and serve with very nicely poached eggs on the top of it; garnish with sippets of fried or toasted bread.

Curry of Cold Meat.

- *Ingredients* — Some scraps of cold meat.
- Some stock or gravy.
- Curry powder and flour in the proportion of a dessert-spoonful of each to every half pint of gravy.
- 1 small onion.
- 1 small apple.
- ½ oz. of dripping.
- A few drops of lemon juice.
- Salt.
- Some boiled rice.

Method.—Slice the onion and apple, and fry them in the dripping.

When fried, rub them lightly through a hair sieve.

Mix the curry powder and flour smoothly with the stock.

Stir and cook well over the fire.

Add the onion, apple, lemon juice, and salt.

86 Then lay in the meat, and let it warm through, being careful that the sauce does not boil.

Serve with nicely boiled rice.

Shepherd's Pie.

- *Ingredients*—Slices of cold meat.
- Boiled potatoes.
- Butter or dripping.
- A little gravy.
- Pepper and salt.

Method.—Season the pieces of meat with pepper and salt, and lay them in a pie-dish with a little gravy.

Mash the potatoes smoothly with butter or dripping; and pepper and salt to taste.

Spread the potatoes over the meat in the form of a pie-crust, and smooth them with a knife dipped in hot water.

Bake for half an hour.

Patties.

- *Ingredients*—Some scraps of cold meat.
- A little gravy.
- Pepper and salt.
- Pastry.

- 1 egg.

Method.—Mince the meat and moisten with the gravy, adding pepper and salt to taste.

If veal or chicken are used, mince a little ham with them, and add a few drops of lemon juice.

Roll out the pastry, and stamp it into rounds with a fluted cutter.

Lay half the rounds on greased pattypans.

Brush round the edges of the paste with a little beaten egg, and put a little mince on each round.

Cover them with the remaining rounds of paste, pressing the edges lightly together.

Glaze with the beaten egg, and bake in a quick oven for about 15 minutes.

87

Fritters.

- *Ingredients*—Some cold meat.
- Some nice gravy.
- Some Kromesky batter.

Method.—Cut the meat into neat pieces; dip them in the batter and fry in hot fat until lightly browned (*see* French Frying).

Pile on a hot dish, and serve, if possible, with a nice gravy poured round them.

Rissoles.

- *Ingredients*—Some boiled potatoes.
- Cold meat.

- A little butter.
- 2 eggs.
- Bread-crumbs.
- Pepper and salt.

Method.—Take equal quantities of boiled potatoes and cold meat.

Mash the potatoes with butter, and add the meat finely minced.

Mix this thoroughly with a beaten egg, adding pepper and salt to taste.

Form into balls or egg shapes.

Egg and bread-crumb them, and fry them in hot fat (*see* French Frying).

Dish on a folded napkin, and garnish with fried parsley.

Cold Meat with Purée of Tomatoes.

- *Ingredients*—Slices of cold meat.
- 4 or 5 tomatoes.
- 1 small slice of bacon.
- 1 bay leaf.
- 1 piece of carrot, turnip, and onion.
- 1 88 sprig of parsley.
- Thyme and marjoram.
- 1 teaspoonful of vinegar.
- Pepper and salt.

Method.—Cut the bacon into dice, and fry it.

As soon as the fat melts, put in the tomatoes and other vegetables, cut in slices; stir them, and fry lightly, and then rub through a hair sieve.

Add the vinegar and pepper and salt.

Make the *purée* hot in a saucepan, and lay the pieces of meat in it to warm through.

Serve in a hot dish, with a border of boiled rice or macaroni.

Cold-meat Pie.

- *Ingredients*—Slices of cold meat.
- (If liked, slices of cold boiled potatoes).
- Some stock or gravy.
- Pepper and salt.
- Some plain pastry.

Method.—Roll out the paste, and cut a piece large enough for the cover.

Roll out the scraps, and from them cut a band an inch wide.

Wet the edge of the dish and place this round it.

Season the meat with pepper and salt, and lay the slices in the dish alternately with the potatoes.

Raise them in the middle of the dish in a dome-like shape, and pour in some gravy.

Wet the edges of the band of paste, and lay the cover over.

Trim round neatly, and make a hole in the middle of the crust.

Brush over with beaten egg, and decorate with paste leaves.

Bake in a quick oven for half an hour.

89

Cold Meat and Macaroni.

- *Ingredients*—Slices of cold meat.
- Macaroni.

- Stock.
- Bread-crumbs.
- And, if possible, 2 or 3 tomatoes.

Method.—Put the macaroni in boiling water, and boil it 20 minutes.

Then pour away the water, and stew it in the stock until tender.

Put a layer of macaroni in the bottom of a greased pie-dish.

Lay on it the meat, and cover it with another layer of macaroni, seasoning with pepper and salt.

Proceed in this way, until the dish is full (the top layer must be macaroni).

If tomatoes are used, slice them, and lay over the top; sprinkle with brown crumbs, and bake for about 20 or 30 minutes.

Mayonnaise of Cold Meat.

- *Ingredients*—Slices of cold meat.
- Green salad.
- Beetroot.
- Hard-boiled egg.
- Some Mayonnaise sauce.

Method.—Slice the salad, and mix the meat with it.

Heap it high on a glass or silver dish.

Garnish with beetroot and hard-boiled egg, and pour Mayonnaise sauce over (*see* Sauces).

Beef and Mushrooms.

- *Ingredients*—1½ lb. cold roast beef.

- 1 dozen mushrooms.
- 1 shalot or small onion, very finely chopped.
- 2 oz. 90 of butter.
- ½ pint of beef gravy.
- 1 dessertspoonful of vinegar.
- Pepper and salt to taste.

Method.—Cut the beef into neat slices, and wash and peel the mushrooms.

Season the meat with pepper and salt, and lay half of it in the bottom of a pie-dish.

Place some of the mushrooms on the top of it.

Put 1 oz. of butter, in pieces, about them.

Then put in the remaining pieces of beef, and the mushrooms and butter in the same way.

Pour in the gravy and vinegar, and cover closely.

Put it into a moderate oven to bake for three-quarters of an hour.

Beef Scalloped.

- *Ingredients*—Some cold roast beef minced.
- 1 boiled onion, very finely chopped.
- Some mashed potatoes.
- Butter.
- Pepper and salt.
- 1 egg.
- A little gravy and mushroom catsup.

Method.—Mince the beef finely, and moisten it with a little nice gravy.

Add the onion to it, and season nicely with catsup.

Mix the mashed potatoes with plenty of butter, and the egg well beaten, pepper and salt.

Place the mince in greased scallop shells, and cover with the potatoes.

Bake in a quick oven until lightly browned.

When economy has to be studied, leave out the eggs and substitute clarified dripping for the butter. The mixture can be baked in a pie-dish, if more convenient.

91

Cold Beef Olives.

- *Ingredients* — Some cold roast beef.
- Some veal forcemeat, omitting the suet.
- Some gravy.
- Flour.
- Pepper and salt.
- Some mashed potatoes.

Method. — Take slices of cold beef, and cut them into strips 1½ inches in width.

Lay on each a little veal stuffing; roll them round it, and tie them with string.

Put them into a stewpan close together; pour the gravy over them, and simmer them gently for ten minutes.

Dish them on a border of mashed potatoes.

Thicken the gravy with a little flour, and pour it over them.

92

ENTRÉES.

Quenelles of Veal.

- *Ingredients*—1 lb. of fillet of veal.
- 1 oz. of butter.
- 2 oz. of flour.
- 1 gill of water.
- A few drops of lemon juice.
- 2 eggs.
- Seasoning.

Method.—Scrape the veal finely.

Melt the butter in a saucepan; mix in the flour.

Then add the water and cook well.

Put this panada into a mortar with the veal, eggs, lemon juice, and seasoning, and pound thoroughly.

Then rub through a wire sieve.

Shape the mixture somewhat like eggs with dessertspoons and a knife dipped in hot water.

Poach them gently in a greased frying-pan, or *sauté* pan, for ten minutes.

Dish them on a border of mashed potatoes, and pour white sauce over them.

Garnish with chopped truffle and ham.

Cooked green peas, mushrooms, or other vegetables, may be placed in the centre.

Mutton Cutlets à la Macédoine.

- *Ingredients*—Part of best end of neck of mutton.
- 1 egg.
- Bread-crumbs.
- 3 oz. 93 of clarified butter.
- Seasoning.

Method.—Saw off the chine bone, and the ends of the rib bones, leaving the cutlet bone three inches in length.

Cut the cutlets with a bone to each, and beat them with a cutlet bat to about half an inch in thickness.

Trim them, and leave half an inch of the rib bone bare.

Season, egg and bread-crumb, and fry in clarified butter in a *sauté* pan for five or seven minutes.

Dish on a border of mashed potatoes, put a *macédoine* of vegetables in the centre, and pour brown sauce round them.

Mutton Cutlets à la Rachel.

- *Ingredients*—Some mutton cutlets.
- *Foie gras.*
- Brown sauce.
- *Macédoine* of vegetables.
- Mashed potatoes.
- Truffle.
- Pigs' caul.

Method.—Plainly fry some mutton cutlets, coat one side of each cutlet with the *foie gras*, smoothing it with a knife dipped in hot water.

Lay a small piece of truffle on each cutlet and cover them with pigs' caul.

Put them on a baking-sheet in a moderate oven for about a quarter of an hour.

Dish them on a border of mashed potatoes.

Pour brown sauce round them, and put a *macédoine* of vegetables in the middle.

Epigrammes.

- *Ingredients*—The rib part, which was sawn off the mutton cutlets.
- Egg and bread-crumbs.

94 *Method.*—Boil the mutton until the bones can be easily removed.

Press it, and, when cold, cut it into cutlets or other shapes.

Egg and bread-crumb twice, and fry in hot fat (345°) in a frying-basket.

Dish on a border of mashed potatoes, and pour brown sauce round them.

Any cooked vegetables can be put in the centre for a garnish.

Chicken Croquettes.

- *Ingredients*—2 oz. of cooked chicken.
- 1 oz. of cooked ham.
- 1 oz. of butter.
- ¾ oz. of flour.
- 1 gill of stock.
- ½ gill of cream.
- 6 button mushrooms.
- A few drops of lemon juice.
- Seasoning.

- Pastry.

Method. — Mince the chicken, ham, and mushrooms.

Melt the butter in a small stewpan.

Mix in the flour.

Pour in the stock, and cook well.

Then add cream, lemon juice, and seasoning; lastly, the chicken, ham, and mushrooms.

Spread on a plate to cool.

Roll out some paste as thin as possible.

Cut into rounds.

Put a little of the mixture on each, and egg round the edges.

Fold them over, egg and bread-crumb the *croquettes*, and fry in a frying-basket in hot fat (*see* French Frying).

Garnish with fried parsley.

95

Veal Cutlets à la Talleyrand.

- *Ingredients* — 7 or 8 veal cutlets.
- 1½ oz. butter.
- 3 button mushrooms, chopped.
- 1 small shalot, chopped.
- A teaspoonful of chopped parsley.
- The yolks of 2 eggs.
- 2 tablespoonfuls of cream.
- A few drops of lemon juice.
- 1 gill of white sauce (*see* Sauces).
- Some mashed potatoes.
- A few green peas.
- Pepper and salt.

Method.—Fry the cutlets in the butter, sprinkling the mushroom, shalot, and parsley under and over them.

When the cutlets are cooked, remove them from the pan and pour in the white sauce and cream.

Stir briskly over the fire.

Then add the yolks of the eggs; let them thicken in the sauce, but be careful not to curdle them.

Take the pan off the fire, and add the lemon juice and seasoning as required.

Dish the cutlets on a border of mashed potatoes.

Pour the sauce over them, and put a few nicely cooked peas, or other appropriate vegetables, in the middle.

Fillets of Beef à la Béarnaise.

- *Ingredients*—7 or 8 nice little fillets.
- 1½ oz. of butter.
- Mashed potatoes.
- ½ pint of brown sauce (*see* Sauces), or good gravy.
- Some good *Béarnaise* sauce (*see* Sauces).

Method.—Fry the fillets in the butter.

96 Dish them on a border of mashed potatoes.

Pour brown sauce or gravy round them, and put the *Béarnaise* sauce in the middle of the fillets.

Rabbits à la Tartare.

- *Ingredients*—1 rabbit.
- Some browned bread-crumbs.
- 1 egg.

- ½ pint of Tartare sauce (*see* Sauces).

Method. — Cut the rabbit into joints.

Dry them well.

Egg and bread-crumb them.

Put them on a greased baking-sheet, with pieces of butter on them.

Bake for half an hour, being careful not to dry them up too much.

Pour the sauce on a dish and pile up the rabbit in the middle of it.

Chicken à la Tartare.

Proceed as in the foregoing recipe, substituting a chicken for a rabbit.

Pigeons Stewed à l'Italienne.

- *Ingredients* — 3 pigeons.
- 1 piece of carrot, turnip, and onion.
- 1 pint of stock.
- 1 sprig of parsley, thyme, and marjoram.
- 1 bay leaf.
- If possible, 1 or 2 tomatoes.
- 1 wineglass of sherry.
- 2 oz. of butter.
- 1 oz. of flour.
- Some mashed potatoes.
- A *macédoine* of vegetables.

Method. — Have the pigeons trussed as for stewing.

97 Cut them in two, and fry them in the butter.

Then remove the pigeons, and fry the vegetables.

Stir the flour, and when that is a little brown, pour in the stock or sherry. Put in the pigeons and stew gently until they are tender.

Dish them in a circle on a border of mashed potatoes.

Strain the gravy over, and put a *macédoine* of vegetables in the centre.

Croustards à la Reine.

- *Ingredients*—Some puff pastry.
- A little *quenelle* meat (*see* Quenelles of Veal).
- 1 gill of white sauce.
- 3 oz. of cold chicken minced.
- 1 oz. of cooked ham minced.
- 2 or 3 button mushrooms finely chopped.
- 2 tablespoonfuls of cream.
- A little thick white sauce.
- Ham or truffle for decoration.

Method.—Line some little tartlet tins with some puff paste, put a piece of dough in each, and bake them.

Mix the chicken, ham, and mushrooms with the white sauce and cream. Add pepper and salt to taste.

Remove the paste cases from the tins, take the dough from the middle, and fill them with the chicken mixture.

Cover the top of each with the *quenelle* meat spread like butter, put them into the oven for a few minutes to cook the *quenelle* meat.

When dishing them up, spread a little thick white sauce on the top of each, and ornament them with ham and truffle.

Sweetbreads à la Béchamel.

- *Ingredients*—1 dozen lambs' heart sweetbreads.

- 1¼ pint of veal stock.
- 1 oz. of <u>butter.</u>
- 1 oz. 98 of flour.
- A small piece of carrot, turnip, and onion.
- 1 sprig of parsley.
- 2 tablespoonfuls of cream.
- A slice of lean ham.
- A few drops of lemon juice.
- Some mashed potatoes.
- A few green peas nicely boiled.
- A little finely-chopped cooked ham.
- Some parsley or truffle.
- Pepper and salt.

Method.—Trim the sweetbreads, and soak them in cold water for two hours.

Then throw them into boiling water, and simmer them gently for five minutes.

Soak them again in cold water for twenty minutes.

Then put them in a stewpan with the stock, carrot, turnip, onion, parsley, and ham.

Simmer gently until the sweetbreads are quite tender.

Then remove them, and add to the stock the flour mixed thoroughly with butter.

Stir and boil well, to cook the flour.

Add the cream, lemon juice, and seasoning.

Strain the sauce through a fine strainer or tammy-cloth.

Dish the sweetbreads in a circle on a border of mashed potatoes.

Pour the sauce over them.

Put on each sweetbread a tiny pinch of finely-chopped parsley, ham, or truffle; or use all three, placing them alternately.

The green peas should be put in the centre of the dish.

Braised Sweetbreads.

- *Ingredients*—2 calves' sweetbreads.
- 1 pint of strong second stock.
- A 99 piece of carrot, turnip, and onion.
- 1 sprig of parsley, thyme, and marjoram.
- 1 bay leaf.
- Some larding bacon.
- Some carrots and turnips cut in fancy shapes.

Method.—Soak the sweetbreads in cold water for quite two hours.

Then put them in boiling water, and simmer them for ten minutes to make them firm.

Soak them again in cold water for twenty minutes, and then lard them nicely.

Put the vegetables, cut in pieces, in the bottom of a stewpan.

Lay the sweetbreads on them, and pour in the stock; it should come half way up the sweetbreads.

Cover them with buttered paper, and put the lid on the stewpan.

Simmer gently until the sweetbreads are tender.

Then put them on a baking-tin, and put them in the oven to brown.

Strain the stock they were cooked in into a large saucepan, and boil it rapidly down to a glaze.

Put the sweetbreads on a hot dish, and pour the glaze over.

Carrots and turnips may be cut in fancy shapes, and nicely boiled to garnish the dish.

If preferred, the sweetbreads can be cooked without being larded; a slice of very thin bacon being laid on the top of each.

If a proper braising-pan is used, the sweetbreads need not be browned in the oven.

Lambs' sweetbreads can be cooked the same way. One dozen will be wanted for a small dish.

100

Sweetbreads à la Parisienne.

- *Ingredients*—1 dozen lambs' heart sweetbreads.
- 1 pint of good second stock.
- 2 oz. of butter.
- 1 oz. of flour.
- A piece of carrot, turnip, and onion.
- 1 sprig of parsley.
- 1 dessertspoonful of mushroom catsup.
- 1 wineglass of sherry.
- A few drops of lemon juice.
- Pepper and salt.
- Some mashed potatoes.
- Green peas nicely cooked.

Method.—Trim the sweetbreads and soak them for two hours; throw them in boiling water, and simmer them gently for five minutes; then soak them in cold water for twenty minutes.

Simmer them in the stock until they are quite tender.

Then make the butter quite hot in a stewpan.

Fry the sweetbreads in it until nicely browned.

Remove them and fry the flour; then pour in the stock, and stir, and cook well; add the catsup, wine, and lemon juice.

Dish the sweetbreads on a border of mashed potatoes, and pour the same over them.

Put a garnish of nicely cooked green peas in the middle.

Minced Sweetbread.

- *Ingredients* — The remains of dressed sweetbreads.
- 2 or 3 mushrooms.
- Enough stock to moisten nicely.
- 1 teaspoonful of flour.
- A slice of cooked ham.
- A few drops of lemon juice.
- 1 oz. of butter.
- Pepper and salt.

101 *Method.* — Mince the sweetbreads, mushrooms, and ham.

Melt the butter in a stewpan, and fry the mushrooms in it.

Put in the flour, and mix it smoothly with the butter.

Then put in the sweetbread and ham, and enough stock to mix nicely.

Add lemon juice, pepper, and salt, to taste.

Make it hot, and then put the mixture into oiled-paper cases.

Sprinkle over the top of each a few browned crumbs and put in the oven for a few minutes.

Fried Sweetbread.

- *Ingredients* — 1 dozen lambs' heart sweetbreads.
- 1 pint of good stock.
- 1 oz. of butter.
- 1 oz. of flour.
- A few drops of lemon juice.
- If liked, ½ wineglass of sherry.
- Eggs and bread-crumbs.
- Some mashed potatoes and green peas.

Method. — Trim the sweetbreads, and soak them in cold water for two hours.

Then throw them into the boiling stock, and simmer them for half an hour or more until quite tender.

If possible, let them get cold in the stock.

Then egg and bread-crumb them, and fry them in a frying basket in hot fat (*see* French Frying).

To make a sauce, melt the butter in a saucepan.

Mix in the flour smoothly, pour in the stock, and stir and cook well; add lemon juice, pepper, and salt to taste, and, if liked, a little sherry.

Dish the sweetbreads on the potatoes; pour the sauce round them, and put the peas in the centre.

The sauce should be made before the sweetbreads are fried, that there may be no delay in serving.

102 If calves' sweetbreads are used, proceed in the same way, cutting them in neat slices before frying.

Cutlets of Veal with Tomato Sauce.

- *Ingredients* — 2 lb. of fillet of veal.
- 2 or 3 oz. of butter, or some of the fat skimmed from the stock-pot.
- 1 pint of tomato sauce.
- ¼ lb. macaroni, nicely stewed in milk and seasoned with Parmesan cheese.
- Some mashed potatoes.
- 1 uncooked tomato.

Method. — Cut the veal into neat little cutlets, and fry them nicely in the butter or skimming.

Dish them in a circle on a border of potatoes.

Pile the macaroni high in the middle.

Pour tomato sauce round, and garnish the macaroni with small strips of uncooked tomato.

Beef Olives.

- *Ingredients* — 1½ lb. of thick beefsteak.
- Some veal stuffing.
- 1½ pint of stock.
- 1 oz. of butter.
- 1 oz. of flour.
- A few drops of lemon juice.
- Pepper and salt.
- Some mashed potatoes.
- A few carrots and turnips, cut in fancy shapes, and nicely cooked.

Method. — Cut the beef into thin strips, lay a little forcemeat on each, and roll them up.

Tie each roll with a little fine string.

Put them in a stewpan close together, and cover them with the stock.

103 Stew them gently for two or three hours until quite tender.

Then place them in a circle on a border of mashed potatoes.

Remove any fat from the stock, and stir in the butter and flour thoroughly mixed together.

Cook the flour well, and then add the lemon juice and seasoning.

Strain the sauce over the olives, and put the vegetables in the centre.

Veal à la Béchamel.

- *Ingredients* — 1 lb. of cold cooked veal.
- ¼ lb. of button mushrooms.
- ½ pint of *Béchamel* sauce.
- The yolks of 2 eggs.
- Some fried sippets of bread.

Method. — Cut the veal into large dice.

Clean the mushrooms and stew them in the sauce until tender.

Then add the yolks of two eggs well beaten.

Stir over the fire until they thicken, but on no account let the sauce *boil*, as that might curdle the eggs.

Last of all, put in the pieces of veal, and let the saucepan remain by the fire until they are thoroughly heated.

Serve garnished with fried sippets of bread.

Grenadines of Veal.

- *Ingredients* — 2 lb. of veal.
- Some larding bacon.
- Some good second stock.
- 1 piece of carrot, turnip, onion.
- 1 sprig of parsley, thyme, and marjoram.
- Some nicely boiled green peas.

Method. — Cut the fillet into nice oval-shaped cutlets, about half an inch in thickness, and lard them.

104 Put the vegetables, cut in small pieces, at the bottom of the stewpan.

Lay the cutlets on them, and pour in sufficient stock to come half way up the cutlets.

Cover them with buttered paper, and put them on a slow fire to simmer gently until tender.

Then put them on a baking-tin in the oven to brown.

Strain the stock and boil it with a half-pint more to a strong glaze.

Dish the *grenadines* on a border of mashed potatoes.

Pour a little glaze over each, and put the green peas in the middle.

Mayonnaise of Fowl.

Cold Entrée for Suppers.

- *Ingredients* — 2 fowls.
- ½ pint of *mayonnaise* sauce.
- A cucumber.
- 4 hard-boiled eggs.
- 1 pint of aspic jelly.
- A beetroot.

Method. — Boil the fowls and cut them into neat joints.

Put them in a dish in a circle, the one leaning on the other.

Place in the middle a bunch of endive, and coat the pieces of chicken with *mayonnaise* sauce.

Cut the hard-boiled eggs in quarters, and lay them round the chicken with slices of cucumber and beetroot, and garnish with a border of chopped aspic.

Veal Cutlets.

- *Ingredients* — 2 lb. of veal cutlet.
- Egg and bread-crumbs.
- 3 oz. of clarified butter.

- ½ oz. of flour.
- ½ pint of nice stock.
- Some mashed potatoes.

105 *Method.*—Beat the cutlet well to break the fibre of the meat, and then cut it into neat oval or round shapes.

Brush them with the egg and cover them with fine bread-crumbs.

Fry them in a cutlet-pan in the butter.

When they are cooked pour some of the butter from the pan.

Stir in the flour smoothly.

Pour in the stock, and cook well.

Add pepper and salt and a few drops of lemon juice.

Dish the cutlets in a circle on a border of mashed potatoes.

Strain the gravy round them, and put some nice little rolls of bacon in the middle.

To cook the bacon, cut it in thin slices; roll them, and put them on a skewer, they may be either toasted or baked.

Veal Cutlets à l'Italienne.

- *Ingredients*—1½ lb. of fillet of veal.
- Cut into neat cutlets.
- 2 oz. of butter.
- Egg and bread-crumbs.
- Some carrot and turnip, cut in fancy shapes and boiled.
- ½ pint of Italian sauce.

Method.—Egg and bread-crumb the cutlets and fry them in the butter.

Dish them on a border of mashed potatoes.

Pour Italian sauce over, and put the vegetables in the middle.

Make the Italian sauce with the butter the cutlets are fried in.

Fillets of Chicken.

- *Ingredients*—Some little fillets of chicken cut from the breast.
- Some streaky bacon.
- ½ 106 pint of *Béchamel* sauce, made with white stock.
- Some mashed potatoes.

Method.—Lay the fillet on a greased baking-tin.

Cover with buttered paper and put them into a moderate oven for ten or fifteen minutes.

Dish them on a border of mashed potatoes.

Pour the sauce over and put little rolls of nicely cooked bacon in the middle.

To cook the bacon, cut it into very thin strips and roll them, run a skewer through, and toast them before the fire.

Chicken à la Marengo.

- *Ingredients*—1 chicken.
- 1½ pint of second stock.
- 3 tomatoes.
- 1 piece of carrot, turnip, and onion.
- 1 sprig of parsley, thyme, marjoram.
- 2 oz. of butter.
- 1 oz. of flour.
- A few drops of lemon juice.

Method.—Cut the chicken into neat joints and fry them in the butter.

Then remove them and fry the vegetables.

Add the flour and fry that.

Then pour in the stock; stir and boil for three minutes.

Then put in the chicken and the tomato, sliced.

Simmer for about thirty minutes, until the chicken is quite tender.

Then put the chicken on to an *entrée* dish.

Add some lemon juice to the gravy, and strain over it.

Chicken à la Cardinal.

- *Ingredients*—1 chicken.
- 1½ pint of *Béchamel* sauce.
- 4 ripe tomatoes.

107 *Method.*—Cut the chicken into joints and put them in a stew-pan with the sauce and tomatoes, sliced.

Simmer gently until the chicken is quite tender.

Then place them on a hot *entrée* dish and strain the sauce over them.

Kidneys and Mushrooms.

- *Ingredients*—2 dozen medium sized mushrooms.
- 6 sheep's kidneys.
- 1 pint of second stock.
- 1 oz. of butter.
- 1 oz. of flour.
- 2 tablespoonfuls of cream.
- A few drops of lemon juice.

Method.—Peel the mushrooms, cut off the stalks, and wash them.

Wipe the kidneys and slice them, put them in a stewpan with the stock and mushrooms.

Simmer them gently for thirty minutes or more, until quite tender.

Mix the butter and flour very smoothly, stir them in and boil for about three minutes.

Add the cream and let it boil, season to taste, and squeeze in a few drops of lemon juice.

Curried Rabbit.

- *Ingredients*—1 apple.
- 1 onion.
- 2 dessertspoonfuls of curry powder.
- 1½ pint of second stock.
- 2 tablespoonfuls of cream.
- 2 oz. of butter.
- 2 dessertspoonfuls of flour.
- Salt.
- A few drops of lemon juice.

Method.—Cut the rabbit into neat joints and fry them in the butter.

108 Then remove them and fry the onion and apple, sliced.

Mix the curry powder and flour smoothly with the stock.

Put it into a stewpan; stir and boil three minutes.

Put in the rabbit and add the onion and apple, which should be rubbed through a hair sieve.

Simmer gently for thirty minutes or more, until the rabbit is tender.

Add the cream and let it boil in the sauce.

Squeeze in the lemon juice and add salt.

If a dry curry is liked, remove the rabbit when tender, and boil and reduce the sauce to half the quantity, leaving only sufficient to coat the pieces of rabbit well.

Serve nicely cooked rice with the curry (*see* Rice for Curry).

Curried Chicken.

Make according to the directions in the preceding recipe, using white stock or boiled milk.

Mutton Cutlets à la Milanaise.

- *Ingredients*—7 or more mutton cutlets.
- 2 eggs, white bread-crumbs.
- 3 oz. Parmesan cheese, grated.
- A little boiled macaroni.
- ½ pint brown sauce.
- Some mashed potatoes.
- 2 oz. clarified butter, or the fat skimming of the stock-pot.

Method.—Trim the cutlets neatly.

Brush them with egg and cover them with bread-crumbs mixed with 2 oz. of the grated cheese.

Fry them for about five minutes in a cutlet pan.

Dish them on a border of mashed potatoes and put some nicely-cooked macaroni in the centre with 1 oz. of grated cheese.

Pour the brown sauce round them and serve very hot.

109

Chaud-froid Chicken.

Cold Entrée for Suppers and Luncheons.

- *Ingredients* — The best joints of 2 chickens.
- 1 pint of *Béchamel* sauce.
- ¼ oz. of Swinborne's or Nelson's Gelatine.
- Some aspic jelly.
- Endive and lettuce.

Method. — Melt the gelatine and mix it with the sauce.

Coat the pieces of chicken carefully with it, giving them each two coats if they require it.

When the sauce is firm, place them in a circle on an *entrée* dish.

Put some lettuce, nicely mixed with salad dressing, in the centre, and garnish prettily with the endive.

A border of aspic jelly should be placed round the chicken.

If liked, the chicken may be decorated with truffle or ham.

Rissoles of Game.

- *Ingredients* — Some scraps of cold game.
- Some very stiff second stock.
- Lemon juice, pepper, salt.
- Egg and bread-crumbs.

Method. — Mince the game finely.

Melt the stock and moisten the game well with it.

Add pepper and salt, and a few drops of lemon juice.

Spread the mixture on a plate to get cold.

When cold it will be quite firm.

Mould it into balls or egg shapes.

Cover them with egg and bread-crumbs, and fry them in hot fat (*see* French Frying).

Serve on a folded napkin, and garnish with fried parsley.

110

Podovies.

- *Ingredients* —Some cooked beef, minced finely.
- A little thick gravy, lemon juice.
- A little pastry.
- Pepper and salt.
- Some crushed vermicelli and one or two eggs.

Method. —Mix the beef with the gravy; season it with pepper and salt.

Roll out the pastry as thin as possible.

Cut it into rounds with a good-sized cutter.

Brush the edges of the rounds with beaten egg, and put a little of the minced meat in the middle of each.

Fold them over, pressing the edges well together.

Cover with the egg, and then with the vermicelli.

Drop them into hot fat (*see* French Frying) and fry them a golden brown. As they will rise to the top of the fat, it will be necessary to keep them under with a wire basket or spoon. Dish on a folded napkin and garnish with fried parsley.

111

FISH COOKERY.

To Boil Fish.

Be very careful that the fish is thoroughly cleansed, then place it on the fish-strainer, and tie a cloth, or piece of muslin, over it. (This is to prevent any scum settling on the fish to disfigure it, or spoil its colour.) Immerse it in boiling water, to which two tablespoonfuls of salt, and two of vinegar, have been added; boil it for three minutes to set the albumen on the outside, and so form a casing to keep in the juices and flavour of the fish. Then draw the kettle to the side of the fire and simmer gently until the fish is cooked. For a thick piece of fish, six minutes to each pound, and six minutes over, is the time usually allowed; but no hard-and-fast rule can be laid down, as the time it will take to cook depends on the size and shape, as well as on the weight of the fish. When the fish is cooked, it will have an opaque appearance; and on being pulled, will leave the bone readily. Care must be taken to cook it sufficiently but not to over-boil it. Under-done fish is very unpleasant, while over-cooked fish is flavourless, and breaks to pieces.

Salt fish is put into lukewarm water for the purpose of drawing out some of the salt, and must be simmered until tender. Mackerel should also be put into lukewarm water, as the skin is very tender, and boiling water would break it.

When the fish is cooked, remove the cloth, or muslin, and place the strainer across the kettle that the fish may get well drained. Cover it with a hot cover, and leave it in that position for a few minutes. Then dish, on a folded 112 napkin; or on a strainer, if sauce is poured over it. Garnish tastefully, and serve with an appropriate sauce. Small cod, or salmon, if boiled whole, should be trussed in the form of the letter S.

Baked Fish.

The oven should be kept at a moderate heat, that the fish may not be dried up. Small fish may be cooked with great advantage in the oven, if carefully covered with buttered paper, which will keep them moist, and prevent any baked flavour.

Fried Fish.

Small fish, such as whiting, smelts, &c., are generally fried whole. Larger fish, such as cod and salmon, are fried in the shape of cutlets. Fish to be fried, must be covered with egg and crumbs, or batter. A stewpan, half full of fat, and not a frying-pan, should be used for the purpose (*see* French Frying), except in the case of the sole; and for that, the new fish-fryer, with a wire strainer, is far better than the old-fashioned pan. The bread-crumbs, for fish, should be prepared by rubbing stale bread through a wire sieve.

Boiled Turbot.

Boil it according to the directions for boiling fish. It usually takes from half an hour to an hour, according to its size. It should be dished on a folded napkin, with the white side uppermost; and garnished with cut lemon, parsley, and coral. Serve with it lobster, shrimp, or anchovy sauce.

Boiled Brill.

This fish is cooked like turbot; garnished in the same way, and served with the same sauces.

113

Boiled Salmon.

Boil according to the directions given for boiling fish. Truss a small salmon in the form of the letter S. Dish on a folded napkin; and garnish with parsley and coral. Serve with lobster, shrimp, anchovy, or tartare sauce.

Boiled Cod.

Boil according to directions given for boiling fish. A small piece is often served with thick egg-sauce poured over it, and garnished with the yolk of an egg rubbed through a wire sieve.

Salt Cod, Haddock, Plaice, and any Fish,

May be boiled according to directions given for boiling fish, and served with egg, anchovy, or any other appropriate sauce.

Curried Fish.

- *Ingredients* — 1½ lb. of cold boiled fish.
- 1 small onion.
- 1 small apple.
- ½ pint of second stock.
- A few drops of lemon juice.
- 1 oz. of butter.
- 1 dessertspoonful of curry powder.
- 1 dessertspoonful of flour.
- Salt.

Method. — Slice the onion and apple; fry them in the butter, and then rub them through a hair sieve.

Mix the flour and curry powder smoothly with the stock.

Stir over the fire and boil well.

Then add the onion, apple, lemon juice, and salt.

Break the fish into pieces, and remove the bones.

114 Put it into the sauce, and let it warm through.

Serve with a border of rice round it.

Kedgeree.

- *Ingredients* — The remains of cooked fish.
- An equal quantity of boiled rice.
- 2 hard-boiled eggs.
- A little butter.
- Pepper and salt.

Method. — Break the fish into flakes, removing all the bones.

Melt a little butter in a saucepan.

Put in the rice, fish, and the whites of the eggs cut small, pepper and salt.

Stir over the fire until quite hot.

Heap it on a hot dish in the form of a pyramid, and sprinkle over it the yolks of the eggs, rubbed through a wire sieve.

Baked Herrings.

- *Ingredients* — A few herrings.
- Browned bread-crumbs.
- A little butter or dripping.
- Parsley.

Method. — Split open the herrings, and remove the back-bone.

Roll them up, and place them with their roes on a greased baking-sheet.

Cover them with greased paper, and put them into a moderate oven for ten or fifteen minutes until cooked.

Place the rolls on a folded napkin, and sprinkle some brown bread-crumbs in a straight line on each.

Garnish with the roes and sprigs of parsley.

115

Herrings baked in Vinegar.

- *Ingredients* — A few herrings.
- 1 dessertspoonful of finely-chopped parsley.
- 1 small onion.
- Vinegar.
- Pepper and salt.

Method. — Grease a pie-dish, and put some herrings at the bottom.

Sprinkle them with the parsley and onion finely chopped, and the pepper and salt.

Put another layer of herrings on the top, and sprinkle them similarly.

Proceed in the same way until the dish is full.

Cover them with vinegar.

Place over them a dish, and bake in a slow oven for three or four hours.

Herrings cooked in this way are used cold.

Smelts Fried.

- *Ingredients* — Smelts.
- Egg.
- Bread-crumbs.

- Parsley.

Method. — Dry the smelts well, and fix their tails in their mouths.

Cover them with egg and bread-crumbs, and fry them a golden brown in a frying-basket in hot fat (*see* French Frying).

Garnish with fried parsley, and serve with melted butter or other suitable sauce.

116

Smelts au gratin.

- *Ingredients* — Some smelts.
- A few button mushrooms.
- 1 shalot.
- 1 sprig of parsley.
- Lemon juice.
- Pepper and salt.
- Browned bread crumbs.
- Glaze.

Method. — Lay the smelts on a greased baking-sheet.

Sprinkle under and over them the parsley, shalot, and mushrooms, finely chopped, with lemon juice, pepper, and salt.

Cover them with browned bread-crumbs, and put little bits of butter over them.

Bake them in a moderate oven for seven or ten minutes. Put them on a hot dish, and pour melted glaze over them.

Ling and Hake.

These two fish may be cooked according to any of the recipes given for dressing cod.

Salmon à la Tartare.

- *Ingredients* — A piece of salmon.
- Some tartare sauce.
- Chopped parsley.
- Coral.

Method. — Boil the salmon carefully according to the directions given for boiling fish.

Garnish with coral and parsley, and serve with tartare sauce (*see* Sauces).

If the salmon is served cold, the tartare sauce is poured over it. If hot, it is served in a sauce-boat.

117 A slice of salmon is frequently grilled, and served with tartare sauce.

Pickled Salmon.

- *Ingredients* — Some boiled salmon.
- 1 dozen peppercorns.
- 2 saltspoonfuls of salt.
- 3 bay leaves.
- Equal quantities of vinegar and the liquor the fish was boiled in.

Method. — Lay the salmon in a deep pan or pie-dish.

Boil the fish liquor, vinegar, and other ingredients for a quarter of an hour.

Let it get cold, and then pour over the salmon, which should be allowed to remain in the pickle until the next day.

Whitebait.

- *Ingredients*—Whitebait.
- Flour.

Method.—Put plenty of oil or fat into a stewpan, and make it hot (*see* French Frying). The heat of the fat for whitebait should be 400°.

Have a good heap of flour on a cloth.

As soon as the fat is hot, throw the whitebait into the flour, and, taking the cloth by each end, shake the whitebait rapidly until they are well floured.

Turn them quickly into a frying-basket.

Shake the basket well for the loose flour to drop off, and throw the whitebait into the fat for a minute.

As soon as they rise to the surface, remove them with a fish-slice, and drain them on kitchen paper.

Serve them with brown bread and butter, and slices of lemon.

118

Oyster Patties.

- *Ingredients*—6 patty cases.
- 2½ dozen oysters.
- 1 oz. of butter.
- 1 oz. of flour.
- ¼ pint of milk.
- ¼ pint of cream.
- A few drops of lemon juice.
- Pepper and salt.
- A little cayenne.

Method.—Beard the oysters, and cut off the hard white part; cut each oyster in two.

Strain the oyster liquor through muslin.

Put the beards into the milk, and simmer them in it to extract the flavour.

Then melt the butter in a saucepan, and mix in the flour smoothly.

Strain in the milk, and add the oyster liquor. Stir and cook well.

Then add the cream, and let it boil in the sauce.

Lastly, add the pepper, salt, cayenne, and the oysters.

Fill the patty cases with the mixture.

Put the lid on each, and decorate with powdered lobster coral.

Serve hot or cold.

Scalloped Oysters.

- *Ingredients*—Some oysters.
- A little butter, and bread-crumbs.

Method.—Grease some scallop shells, and place on each two or three oysters.

Cover them with broad-crumbs, and put a little piece of butter on each.

Brown them in a quick oven, and serve very hot.

119

Scalloped Oysters à la Française.

- *Ingredients*—2½ dozen oysters.
- 1 oz. of butter.
- 1 oz. of flour.
- ¼ pint of milk.

- 2 tablespoonfuls of cream.
- A few drops of lemon juice.
- Pepper and salt.
- Some bread crumbs.

Method. — Beard the oysters, and cut them in two.

Strain the oyster liquor through muslin.

Simmer the beards in the milk.

Melt the butter in a small stewpan, and mix in the flour smoothly.

Strain in the milk, add the oyster liquor, stir, and cook well.

Then add the cream, and let it boil in the sauce.

Lastly, add lemon juice, pepper, salt, cayenne, and oysters.

Grease some scallop shells, and sprinkle them with bread-crumbs.

Fill them with the mixture, and sprinkle some more crumbs over them.

Brown in a quick oven.

Serve on a folded napkin, and garnish with parsley and cut lemon.

Mackerel à la Normande.

- *Ingredients* — 1 dessertspoonful of bread-crumbs.
- 2 mackerel.
- Half a shalot, chopped finely.
- 1 teaspoonful of finely-chopped parsley.
- ¼ teaspoonful dried and powdered herbs.
- ¼ oz. of butter or dripping.
- Pepper and salt.

120 *Method.* — Split open the mackerel, and remove the back-bones as cleanly as possible.

Grease a baking-tin, and lay one of the mackerel, skin downwards, on it.

Mix the herbs, parsley, shalot, and bread-crumbs together with pepper and salt, and sprinkle them over the fish.

Lay the other mackerel on the top, with the skin uppermost.

Put little bits of butter or dripping about it, and bake from ten minutes to a quarter of an hour.

For serving, sprinkle over a few brown bread-crumbs.

Haddock Stuffed.

- *Ingredients*—1 haddock.
- 3 tablespoonfuls of bread-crumbs.
- 1 dessertspoonful of finely-chopped parsley.
- 1 teaspoonful of dried and powdered herbs.
- Pepper and salt.
- Part of an egg, or a little milk, to bind the stuffing.

Method.—Mix the crumbs, parsley, herbs, pepper and salt, with the egg or milk.

Put the stuffing in the haddock, and fasten it with a small skewer.

Then truss it with string, or two skewers, in the form of the letter S.

Place it on a greased baking-tin; and put a few pieces of butter or dripping on it.

Bake it in a moderate oven for about twenty minutes.

To serve, place it on a dish and remove the skewers.

Garnish with parsley.

121

Cutlets of Cod.

- *Ingredients*—The tail of a cod.
- 1 egg.
- Bread-crumbs.
- Pepper and salt.

Method.—Cut the tail of a cod into neat cutlets.

Season them with pepper and salt, and cover them with egg and bread-crumbs.

Fry them in a frying-basket in hot fat (*see* French Frying).

Serve on a folded napkin, and garnish with fried parsley.

Cutlets of Cod à l'Italienne.

- *Ingredients*—The tail of a cod.
- A little butter.
- Lemon juice.
- Pepper and salt.
- Some Italian sauce.

Method.—Divide the cod into neat cutlets.

Place them on a greased baking-sheet.

Sprinkle over them a few drops of lemon juice, pepper, and salt, and cover them with buttered paper.

Bake them in a moderate oven from ten to twelve minutes.

Dish them in a circle, and pour over them some Italian sauce (*see* Sauces).

Garnish with coral and truffle.

Cutlets of Cod à la Genoise.

Cook some cod cutlets as in preceding recipe, and serve with Genoise sauce (*see* Sauces). Garnish with coral and truffle.

122

Cod with Tomatoes.

- *Ingredients*—1½ lb. of cod cutlets.
- 5 or 6 tomatoes.
- 1 tablespoonful of vinegar.
- Cayenne pepper and salt.

Method.—Rub the tomatoes through a hair sieve.

Then put the *purée* thus obtained into a saucepan, and lay the pieces of cod in it. There should be enough tomato *purée* to cover the cod.

Simmer gently until the cod is tender.

Add the vinegar and seasoning, dish in a circle, and pour the tomato over.

Cod Fricassee.

- *Ingredients*—Some boiled cod, either hot or cold.
- Plain white sauce (*see* Sauces).
- 2 hard-boiled eggs.

Method.—Break the fish into flakes.

Make the sauce quite hot.

Put the fish into it, and warm it through.

There should be just enough sauce to moisten the cod.

Heap it in a pyramid shape on a hot dish.

Garnish it with rings cut from the hard-boiled eggs.

Sprinkle over the top of the cod the yolks rubbed through a wire sieve or strainer.

Cod Sounds Boiled.

- *Ingredients*—Some cod sounds; milk; water.
- Béchamel sauce.
- Pepper and salt.

Method.—Soak the sounds in water for about six hours.

Then boil them in milk and water for half an hour or more until quite tender.

123 Cut them in pieces about two inches square, and make them hot in some *Béchamel* sauce.

Pile them on a dish in the form of a pyramid, with slices of hard-boiled egg, cut lemon, and parsley.

Marinaded Cod Sounds.

- *Ingredients*—Cod sounds.
- Milk.
- Water.
- Oil.
- Vinegar.
- Shalot.
- Parsley.
- Pepper and salt.
- Butter.

Method.—Soak the cod sounds in water for about six hours, and then boil them in milk and water until tender.

Cut them in pieces an inch and a half square.

Mix together equal quantities of oil and vinegar, and add to them a shalot and some parsley, very finely chopped; pepper, and salt.

Steep the sounds in the *marinade*.

Just before serving, dip each one in *Kromesky* batter, and fry in hot fat (*see* French Frying).

Dish in a circle, and pour over them some piquant sauce.

Decorate with truffle and coral.

Cod Stuffed and Baked.

- *Ingredients* — A thick slice from the middle of the cod.
- Some veal stuffing.
- Browned bread-crumbs.

Method. — Fasten the stuffing securely in the cod.

Place it on a greased baking-sheet, and cover it with browned crumbs.

124 Place small pieces of butter or dripping about it, and bake it in a moderate oven for about half an hour, basting occasionally.

Serve with cut lemon, and garnish with parsley.

Note. — A small cod may be stuffed and cooked like a haddock.

Plaice.

This fish may be boiled, baked, or fried.

Fried Fillets of Plaice.

Fillet the plaice by cutting down the centre of the fish with a sharp knife and removing the flesh from either side.

Egg and bread-crumb the fillets, and fry in hot fat (*see* French Frying).

Drain on kitchen paper, serve on a folded napkin, and garnish with fried parsley.

Fried Fillets of Sole.

Prepare like the fillets of plaice, with the exception that the sole should be skinned before it is filleted.

Fish Croquettes.

- *Ingredients* — 1 lb. of cooked fish (haddock, cod, ling, or hake are the best for the purpose).
- 1 oz. of butter.
- 2 or 3 eggs.
- Pepper and salt.
- Some white crumbs.
- Parsley.
- 1 lb. of boiled potatoes.

Method. — Rub the potatoes through a sieve.

Break the fish into flakes, removing the bones.

Mix the fish and potatoes together; blend them 125 thoroughly with the butter, pepper, salt, and a well-beaten egg.

Form the mixture into balls or cakes.

Egg and bread-crumb them, and fry them in hot fat (*see* French Frying).

Serve on a folded napkin, and garnish with fried parsley.

Fish Pudding.

Make a mixture of fish and potatoes as in preceding recipe. Put it on a dish that will stand the heat of the oven, and mould it into the form of a fish.

Bake for half an hour.

Halibut.

This fish may be cooked and served like cod or turbot.

Red Mullets à l'Italienne.

- *Ingredients*—4 or 6 red mullets.
- 2 dessertspoonfuls of mushroom catsup.
- A little butter.
- Lemon juice.
- Pepper and salt.
- Some Italian sauce.

Method.—Lay the mullets in a well-buttered baking-sheet; moisten them with the catsup, and sprinkle with lemon juice, pepper, and salt.

Put some little bits of butter over them.

Bake in a moderate oven for a quarter of an hour or more until cooked.

Lay them on a hot dish.

Mix the liquor from the mullets with some Italian sauce (*see* Sauces), and pour over.

Garnish with truffle and coral.

126

Red Mullets à la Genoise.

- *Ingredients* — Red mullets.
- ½ glass of port.
- A few drops of lemon juice.
- Pepper.
- Some Genoise sauce.
- A little butter.

Method. — Lay the mullets on a well-greased baking-sheet.

Moisten them with the port wine and lemon juice, and put little bits of butter about them.

Bake them in a moderate oven until cooked.

Lay them on a hot dish.

Mix the liquor from the mullets with the Genoise sauce, and pour over them.

Red Mullet in Cases.

- *Ingredients* — 4 red mullets.
- 1 dozen button mushrooms.
- 1 dessertspoonful of finely-chopped parsley.
- 2 shalots.
- Lemon-juice.
- Pepper and salt.
- Salad oil.

Method. — Chop the shalots and mushrooms, and mix them with the parsley.

Oil some pieces of foolscap paper.

Lay the mullets on them; sprinkle over them the parsley, mushroom, shalot, lemon juice, pepper and salt.

Fold them in the cases, and cook on a well-greased baking-sheet, in a moderate oven, for about twenty or thirty minutes.

127

Boiled Whiting.

Fasten the tail in the mouth of each whiting, and lay them on a fish strainer.

Put them into boiling water, with salt in it, and cook them gently for five minutes or more.

Dish on a folded napkin, and garnish with parsley, coral, and cut lemon.

Serve with them *maître d'hôtel*, *Béchamel*, Italian, Genoise, or any other suitable sauce.

Fried Whiting.

- *Ingredients*—Some whiting.
- Egg.
- Bread-crumbs.
- Parsley.
- Lemon juice.

Method.—Skin the whiting, and fasten the tail in the mouth.

Dry them well with a cloth.

Egg and bread-crumb them, and fry them in a frying-basket, in hot fat (*see* French Frying).

Drain them on kitchen paper, and dish on a folded napkin.

Garnish with fried parsley and cut lemon.

Béchamel, lobster, shrimp, Italian, Genoise, or any other suitable sauce, may be served with them.

Whiting à l'Italienne.

- *Ingredients* — Whiting.
- Lemon-juice.
- Pepper and salt.
- A little butter.
- Italian sauce.

Method. — Skin and fillet the whiting.

128 Lay the fillets on a well-buttered baking-sheet.

Sprinkle with lemon-juice, pepper and salt, and cover them with buttered paper.

Cook them in a moderate oven, from seven to ten minutes.

Dish in a circle, and pour Italian sauce over.

Garnish with truffle and coral.

Whiting à la Genoise.

Prepare the whiting as in preceding receipt, substituting Genoise for Italian sauce.

Lobster Cutlets.

- *Ingredients* — 1 hen lobster.
- 1½ oz. of butter.
- 1 oz. of flour.
- 1 gill of cold water.
- 2 tablespoonfuls of cream.
- A few drops of lemon juice.
- Cayenne.

- Pepper and salt.
- Some spawn or coral.
- Egg and bread-crumbs.
- Parsley.

Method.—Remove the flesh from the body of the lobster, and cut it up.

Pound the coral in a mortar, with half an ounce of butter, and rub it through a hair sieve. (If spawn is used it need not be pounded.)

Melt 1 oz. of butter in a stewpan.

Mix in the flour; add the water; stir until it thickens.

Then add the coral, and butter, and cook well.

Next the cream, lemon juice, cayenne, pepper, salt, and lastly the chopped lobster.

Spread the mixture on a plate to cool.

When cool, shape into cutlets.

129 Egg and bread-crumb, and fry in hot fat in a frying-basket.

Put a piece of the feeler in each, to represent a bone.

Garnish with fried parsley.

Lobster Cutlets in Aspic.

Shape some of the lobster-cutlet mixture into cutlets.

Roll in dried and powdered coral, and put a piece of feeler in each.

Pour a little aspic jelly into a clean Yorkshire-pudding tin, or frying-pan.

When set, lay the cutlets on it, and pour in, gently, enough aspic to cover them.

When firm, cut them out with a border of aspic to each, and serve on chopped aspic.

Fried Sole.

- *Ingredients* — A sole.
- Egg.
- Bread-crumbs.
- Parsley.

Method. — Remove the dark skin, and notch the other, here and there, with a knife.

Dry the sole well in a floured cloth.

Brush over with egg, and cover with bread-crumbs.

Flatten them on with a broad-bladed knife, and fry the sole a golden brown in hot fat (for heat of fat *see* French Frying).

A fish-fryer, or a deep frying-pan, should be used for the purpose; and there should be sufficient fat to cover the sole, so that it will not require turning.

When cooked, drain on kitchen paper.

Dish on a folded napkin.

Garnish with fried parsley.

130

Sole à la Parisienne.

- *Ingredients* — 1 sole.
- 1 wineglass of sherry.
- ½ pint of good second stock.
- A few drops of lemon juice.
- 1 teaspoonful of Harvey's sauce.
- 1 teaspoonful of anchovy sauce.
- Pepper and salt to taste.

Method.—Remove the dark skin, and notch the other with a knife.

Lay the sole in a baking-pan, and pour over it the stock and sherry.

Cover with a dish, and bake for twenty or thirty minutes in a moderate oven.

Place it on a hot dish.

Boil the stock rapidly down to half the quantity.

Add to it the sauces, lemon juice, and seasoning, and pour it over the sole.

Fillets of Sole à la Rouennaise.

- *Ingredients*—2 or more soles.
- Lemon juice.
- Lobster-cutlet mixture.
- Some white sauce.
- Chopped truffle.

Method.—Remove both skins from the soles, and fillet them.

Spread some of the lobster-cutlet mixture on the half of each fillet, and fold over.

Place on a greased baking-sheet; sprinkle over lemon juice and salt, and cover with buttered paper.

Bake in a moderate oven for about twelve minutes.

Dish in a circle, and pour over white sauce, mixed with chopped truffle.

131

Fillets of Sole à la Maître d'Hôtel.

- *Ingredients* — Fillets of sole.
- Lemon juice and salt.

Method. — Roll or fold the fillets, and cook like the Sole *à la Rouennaise.*

Cover them with the same sauce as in the last recipe, using chopped parsley instead of truffle.

Sole au gratin.

- *Ingredients* — 1 sole.
- 1 dessertspoonful of chopped parsley.
- 1 chopped shalot.
- 6 chopped button mushrooms.
- Lemon juice.
- Pepper and salt.
- ½ oz. of butter.
- Brown bread-crumbs.

Method. — Grease a dish that will stand the heat of the oven.

Sprinkle on it half of the parsley, shalot, and mushroom, with lemon juice, pepper, and salt.

Lay the sole on the mixture, and sprinkle the remainder of the parsley, &c., over it.

Cover with brown bread-crumbs, and put half an ounce of butter about it, in small pieces.

Bake from ten to fifteen minutes, according to size, and serve with glaze poured round it.

Gurnets baked.

- *Ingredients* — 2 or more gurnets.
- Some veal stuffing, omitting the suet.
- A little stock.
- ½ wineglass of sherry.
- 1 132 or 2 dessertspoonfuls of mushroom catsup.
- Some brown sauce.
- Pepper and salt.

Method. — Remove the head and fins of the gurnets, and stuff them with veal stuffing, fastening it in with small skewers.

Lay them on a well-buttered baking-tin, and pour over them the stock, sherry, and catsup.

Bake them in a moderate oven until cooked.

Then place them on a hot dish, mix the liquor from them with the sauce and pour over.

Stewed Eels.

- *Ingredients* — 2 lb. of eels.
- 1 pint of stock.
- 1 wineglass of port.
- 1 tablespoonful of flour.
- A few drops of lemon juice.
- Pepper.
- Salt.
- Cayenne.
- 2 oz. of butter.

Method. — Cut the eels in pieces about 2½ inches long.

Fry them brown in the butter.

Then put them in a stewpan with the stock.

Stew gently, until tender.

Then remove them from the stock, and put them in a hot dish.

Thicken the stock with the flour.

Add the wine, lemon juice, and seasoning.

Pour over the eels, and serve very hot.

133

PASTRY.

Few people are successful in making pastry. Yet, with a little practice, there is no reason why any one should not make it with some degree of perfection, if the following rules are carefully attended to.

Make the pastry in a cool place, not in a hot kitchen. The board, rolling-pin, and hands should be as cold as possible. Handle it very lightly. The colder pastry is kept during making, the lighter it will be, because it will contain more air; cold air occupies a much less space than warm. The colder the air, the greater, consequently, will be its expansion when the pastry is put into a very hot oven. Roll the paste lightly, and not more than necessary. Puff paste is a kind of fine sandwich. There should be a certain number of layers of dough and layers of butter. Take care, therefore, that the butter is not allowed to break through the dough; and be *very careful to follow* the directions given for making this pastry. Its manufacture requires patience, because, if it is not properly cooled between the turns, the friction of rolling will warm the butter, and cause it to smear into the dough. For short crust, rub the butter or fat lightly into the flour with the tips of the fingers; and do not use more water than necessary in mixing it. This is a common mistake; and too much water deprives the paste of its shortness. Short paste is the best for children and persons of weak digestion; the flour in it being more thoroughly incorporated with the fat, gets better cooked. It is, therefore, capable 134 of more perfect mastication than puff or flaky crust, both of which are liable to be swallowed in flakes.

However well pastry is made, success will not be attained unless the oven is rightly heated. The very lightest crusts will often be totally spoiled in the baking because this important point is not attended to. If the oven is not very hot, the fat will melt and run out of the pastry before the starch grains in the flour burst; consequently, they cannot afterwards expand, however hot the oven may be made; and in this way the paste will become heavy. Take great care, therefore, that the oven is very hot when the paste is put into it.

Watch the paste carefully that it does not take too dark a colour. When it is well thrown up and nearly cooked, it may be removed to a more moderately heated part of the oven if it should appear to be browning too quickly.

Ovens in which the heat comes from the bottom are decidedly the best for either cakes or pastry; but no one should expect to bake well in an oven they do not thoroughly understand. There is so much difference in ovens, that the hottest part of one may be the coolest in another. To bake well requires practice and experience, and no one should be discouraged by a few failures.

Puff Paste.

- *Ingredients*—Equal quantities of Vienna flour and butter.
- A few drops of lemon juice.
- Enough water to mix the flour into a nice lithe dough.

Method.—Rub the flour through a wire sieve.

Make a well in the middle, and squeeze in a few drops of lemon juice.

Mix very gradually with very cold water, taking care that the dough is not too stiff.

Then knead and work well about until quite smooth.

135 Set it aside for a few minutes to get quite cold.

Squeeze the butter in a cloth to press out the water.

Roll out the dough, and place the butter, flattened to a third of its size, in the middle.

Then fold the dough from either side over it, pressing the edges together.

Turn it with its edges toward you, and roll out very gently (care must be taken that the butter does not break through the dough).

Fold it again in three, and put it aside to cool for quite a quarter of an hour. The colder it is kept the better.

Then turn its edges towards you, and roll it out again; fold evenly in three, and roll and fold again in the same manner; each roll and fold is called a turn.

Cool the paste for another quarter of an hour.

Then give it two more turns.

Let it cool again; and at the seventh roll it will probably be ready for use.

It is, however, wise to bake a small piece of the paste before using the whole quantity. If the maker has a very light hand it sometimes happens that eight or even nine turns may be necessary to roll the butter sufficiently into the flour.

Patty Cases.

Roll the puff paste, when ready, to rather more than a quarter of an inch in thickness.

Take a fluted cutter about the size of a tumbler.

Dip it in very hot water, and cut the paste into rounds with it.

Mark the middle of these rounds with a cutter about three sizes smaller.

Roll out the remains of the paste to half the thickness of the patties.

Stamp out some rounds for covers with a fluted cutter two sizes smaller than that used for the cases.

136 Put the cases and covers on a baking-tin, and bake in a quick oven for ten or fifteen minutes.

When cooked, lift the lid and scrape out the soft inside carefully.

Good Short Crust.

- *Ingredients*—1 lb. of flour.

- ¾ lb. of butter.
- Enough cold water to mix rather stiffly.
- A pinch of salt.

Method.—Rub the butter into the flour until like fine bread-crumbs.

Mix with cold water, using as little as possible (if too much is used the crust will not be short).

Roll gently to make the paste bind.

If this paste is used for tarts, add one dessertspoonful of castor sugar to the flour.

Plainer Short Crust.

- *Ingredients*—1 lb. of flour.
- ¼ lb. of butter.
- ¼ lb. of lard.
- 1 teaspoonful of baking powder.
- Water enough to bind.

Method.—Make according to directions given in preceding recipe.

Economical Short Crust.

- *Ingredients*—1 lb. of flour.
- ½ lb. of clarified dripping or lard.
- 1 teaspoonful of baking powder.
- Enough water to mix.

To make this crust still plainer, a quarter of a pound only of clarified dripping or lard may be taken, and three good teaspoonfuls of baking powder.

137 *Method.*—Make according to the directions for Short Crust.

Flaky Crust.

- *Ingredients*—1 lb. of flour.
- ½ lb. of butter or dripping.
- A pinch of salt.
- Enough cold water to mix the paste.

Method.—Rub one half of the butter into the flour, as for short crust.

Mix with the water, and roll it out very thinly.

Put the remainder in little pieces on the paste.

Fold in three, and then in three again.

Roll out to the size required.

Rough Puff Paste.

- *Ingredients*—1 lb. of flour.
- ½ lb. of butter, lard, or dripping.
- Salt.
- Cold water.

Method.—Break the fat into the flour in pieces.

Add a pinch of salt.

Mix with a little cold water.

Turn on to a board.

Roll and fold four times.

Flaky Bread Crust.

- *Ingredients*—1 lb. of bread dough.
- Some butter, lard, or dripping.

Method.—Roll out the dough very thin, and spread with the fat.

Fold in two.

Spread again with fat.

Fold in two, and spread once more with fat.

Fold again, and set aside for one hour.

Then roll out and use.

138

Beef-steak Pie.

- *Ingredients*—2 lb. of nice beef-steak.
- ½ lb. of bullock's kidney.
- 1 lb. of flaky or rough puff paste.
- 1 tablespoonful of flour.
- ½ pint of water.
- Pepper and salt.

Method.—Roll the paste to a quarter of an inch in thickness.

Invert the pie-dish, and cut the paste to the size and shape of the under side of it.

Roll out the remainder, and cut a band one inch wide.

Wet the edge of the pie-dish, and place this round it.

Cut the beef into thin strips.

Dip them in flour, and season with pepper and salt.

Roll each of the strips round a tiny piece of fat.

Put them into the pie-dish alternately with pieces of kidney.

Raise them in the middle of the dish in a dome-like form, and pour in the water.

Wet the edges of the paste lining of the dish, and lay the cover over.

Press the edges lightly together, and trim round with a knife.

Make a hole in the middle of the paste to let the gases from the meat escape.

Brush the crust with beaten egg, and decorate with leaves cut from the trimmings.

Bake for about two hours.

The pie should be put into a quick oven until the pastry is cooked; the heat must then be moderated to cook the meat thoroughly without drying up the pastry. If possible, finish cooking the meat on the top of the oven.

Some people prefer stewing the meat before using it in 139 the pie. If this is done, it must be allowed to get cold before the pie is made.

It is an improvement to the pie to put layers of oysters, bearded, alternately with the rolls of beef.

Rabbit Pie.

- *Ingredients* — 1 rabbit.
- ¼ lb. of salt pork.
- 1 lb. of rough puff or flaky paste.
- ½ pint of water.
- 2 hard-boiled eggs.
- Pepper and salt.

Method. — Prepare the paste as for beef-steak pie, lining the dish in the same manner.

Cut the rabbit into neat joints.

Season them with pepper and salt.

Put them in the pie-dish alternately with the pork.

Pour in the water, and cover with the paste.

Brush over with beaten egg, and decorate with paste leaves.

Make a hole in the middle of the crust for the gases to escape.

Bake for about an hour, attending to directions given for baking beef-steak pie.

Mince Pies.

- *Ingredients*—Puff and other pastry.
- Mincemeat.
- Castor sugar.
- White of 1 egg.

The Mincemeat.

- *Ingredients*—1 lb. of suet.
- 1 lb. of apples.
- 1 lb. of sugar.
- 1 lb. 140 of currants.
- 1 lb. of raisins.
- 1 lb. of candied peel.
- The grated rind of 3 lemons.
- ¼ lb. of ratafias soaked in brandy.

Method.—Chop the suet.

Wash and dry the currants.

Stone and cut the raisins in halves.

Peel, core, and mince the apples.

Chop the candied peel.

Mix all the ingredients well together.

Put them into a stone jar; cover closely and keep for a month.

To Make the Pies.

Roll the paste out, and stamp it into rather large rounds with a fluted cutter dipped in hot water.

Lay half the rounds on patty pans.

Wet the edges of the pastry, and put some mincemeat into the middle of each round.

Cover with the remaining rounds, pressing the edges lightly together.

To glaze, brush them with a little white of egg, and dust with castor sugar.

Bake in a quick oven for ten or fifteen minutes.

Mushroom Pie.

- *Ingredients*—Puff, flaky, or short crust.
- Mushrooms.
- Boiled potatoes.
- Butter.
- Pepper and salt.

Method.—Roll out the paste, and prepare a pie-dish as for beefsteak pie.

Mash the potatoes with butter, pepper, and salt.

Peel the mushrooms, and cut off the ends of the stalks.

141 Put the potatoes and mushrooms in alternate layers in the pie-dish.

Cover with the paste, and finish off and decorate like a beef-steak pie.

Bake in a quick oven for about three quarters of an hour.

Pigeon Pie.

- *Ingredients* — 4 pigeons.
- 1 lb. of rump-steak.
- Yolks of 6 hard-boiled eggs.
- Pepper and salt.
- Some puff or other paste made with 1 lb. of flour.

Method. — Prepare the pie-dish, and roll out the paste as for beef-steak pie.

Draw the pigeons, and cut them in halves.

Cut the steak into thin strips, the way of the grain.

Season the steak and pigeons nicely, and put them into the pie-dish with the hard-boiled yolks.

Pour in the water.

Cover with the paste, and finish like a beef-steak pie.

Wash and clean the legs of two of the pigeons, and stick them in the hole in the top of the pie.

Bake for about an hour and a half.

Veal-and-Ham Pie.

- *Ingredients* — 1 lb. of veal cutlet.
- ½ lb. of ham.
- 4 hard-boiled eggs.
- 1 dessertspoonful of chopped parsley.
- 1 lemon.

- Pepper and salt.
- Some puff, flaky, or other pastry, made with 1 lb. of flour.

Method. — Roll out the paste, and prepare the dish as for beef-steak pie.

142 Cut the veal and ham into neat pieces.

Season them well, and sprinkle them with the parsley and lemon juice.

Put them into the pie-dish with the eggs cut in halves.

Pour in the water.

Cover with paste, and decorate like a beef-steak pie.

Bake for about two hours.

Cornish Pasties.

- *Ingredients* — Some plain short crust.
- Equal quantities of beef-steak or beef-skirt and potatoes.
- 1 onion, finely chopped.
- Pepper and salt.

Method. — Cut the meat and potatoes into small dice, and mix them with the onion, pepper, and salt.

Roll out the pastry.

Stamp it into rather large rounds with the lid of a small saucepan.

Wet round the edges of the paste, and place a small heap of meat and potatoes in the middle of each round.

Double the paste, bringing the edges to the top.

Goffer round them with the fingers to form a frill.

Place the pasties on a greased baking-sheet, and bake in a quick oven from half an hour to an hour.

Sausage Rolls.

- *Ingredients*—Some puff or flaky crust.
- Sausages.
- 1 egg.

Method.—Parboil the sausages.

Skin them, cut them in halves, and let them cool.

Roll out the paste; cut it into squares.

Brush the edges with beaten egg.

Lay a half sausage on each piece of paste, and roll the paste round it, pressing the edges together.

143 Brush the rolls with beaten egg.

Lay them on a greased baking-sheet.

Bake in a quick oven for fifteen or twenty minutes.

Apple Tart.

- *Ingredients*—2 lb. of apples.
- 3 oz. of moist sugar.
- Some pastry.
- 5 cloves or the grated rind of a small lemon.
- ¼ pint of water.

Method.—Make some pastry according to directions given for short crust (the quantity made from ¾ lb. of flour will be sufficient).

Roll out the paste in an oval shape to a quarter of an inch in thickness.

Invert a pint pie-dish, lay the paste over it, and cut it the size and shape of the under side of the dish.

Roll out the remaining pieces, and cut in strips about one inch wide.

Wet the edges of the pie-dish, and lay them evenly round it.

Peel, core, and quarter the apples.

Put them into the pie-dish, mixing them with the sugar.

Pile them up well in the middle of the dish, pressing them to an oval shape with the hands.

Pour in the water, and sprinkle over the lemon rind or cloves.

Wet the edges of the pastry, lining the dish, and put over the piece reserved for the cover.

Press the edges lightly together, and trim with a knife.

Make a small hole with a skewer on either side of the cover to let the steam escape.

144 To glaze, brush over with the white of an egg, and dust with castor sugar.

Bake from half to three-quarters of an hour. The oven should be very quick at first, and moderate afterwards.

Any Fruit Tart may be made by this recipe. Sugar must be added according to the acidity of the fruit used.

Genoise Pastry.

- *Ingredients*—6 oz. of flour.
- 6 oz. of butter.
- 8 oz. of castor sugar.
- 7 eggs.

Method.—Melt the butter in a stewpan, and brush over a *sauté* pan or shallow cake tin with it.

Line the pan with paper, and brush that also with the melted butter.

Break the eggs into a basin.

Add to them the sugar, and beat with a whisk for about twenty minutes until they rise.

The basin containing them may be placed on a saucepan of hot water; but care must be taken that the heat is not too great, as that would cook the eggs.

When the eggs are sufficiently beaten, *stir* in the flour and butter *very lightly*.

If *beaten* in, the pastry will not be light.

Pour the mixture into the pan, and bake for about an hour.

Genoise Sandwiches.

- *Ingredients*—Genoise pastry.
- Jam.

Method.—Cut the Genoise pastry into slices.

Spread them with jam.

Lay the slices one on the other, and cut in triangular shapes.

145

Genoise iced-cakes.

- *Ingredients*—Genoise pastry.
- Jam.
- Grated cocoa-nut.
- Iceing.

Method.—Stamp out small cakes of Genoise pastry with a round cutter.

Spread the sides thinly with jam.

Roll the cakes in the cocoa-nut.

Ice round the top of the cakes, and put some jam in the middle of the iceing.

Genoise Preserve-cakes.

- *Ingredients*—Uncooked Genoise pastry.
- Some preserve.
- Some syrup of sugar and water.
- Hundreds and thousands.
- Chopped *pistachio* kernels.
- Grated cocoa-nut.

Method.—Partly fill small well-buttered dariol moulds with the Genoise mixture, and bake in a moderate oven.

When done, and sufficiently cool, cut a small circular piece from the bottom of the cakes.

Scoop out some of the inside, and fill them with the preserve.

Replace the small circular piece.

Brush the cakes over with the syrup, and roll them in the hundreds and thousands, chopped *pistachio*, and cocoa-nut.

They should be entirely covered with the decorations.

Pile them prettily on a dish, and decorate them with holly leaves.

146

Almond Cakes.

- *Ingredients*—Genoise pastry.
- Almonds.
- The white of 1 egg.
- 1 oz. of castor sugar.

Method.—Stamp out the Genoise pastry into small cakes, with round cutters.

Beat the white of egg, mix it with the castor sugar, and spread it over the cakes.

Sprinkle them well with almonds, blanched and chopped.

Put them in a moderate oven to take a pale fawn colour.

Cheese Cakes.

- *Ingredients*—Some remains of puff pastry.
- 2 oz. of sugar.
- 2 oz. of butter.
- 1 lemon.
- Half a sponge cake.
- 1 whole egg and 1 yolk.

Method.—Cream the butter in a basin.

Add to it the castor sugar.

Beat well together, adding one by one the yolks of the eggs.

Then mix in the grated lemon peel, and the lemon juice and the sponge cake, rubbed through a wire sieve.

Lastly, stir in lightly half the white of the egg, beaten to a stiff froth.

Roll out the pastry.

Stamp into rounds with a fluted cutter dipped in hot water.

Lay the rounds in patty pans, and put a little dummy of dough or bread in the middle of each.

Bake them in a quick oven.

147 When nearly cooked, remove the dummies and fill their places with the cheese-cake mixture.

Return them to the oven until the pastry is cooked and the cheese-cake mixture has taken a pale colour.

Tartlets.

- *Ingredients* — The remains of puff paste.
- Some preserve.

Method. — Roll out the paste, and stamp into rounds with a fluted cutter dipped in hot water.

Lay the rounds on patty pans.

Place in the middle of each a dummy, made of dough or bread.

Bake in a quick oven.

When the pastry is cooked remove the dummies, and fill the places with jam.

Plainer tartlets may be made with short, flaky, or other pastry.

Cheese d'Artois.

- *Ingredients* — Remains of puff paste, or some flaky crust.
- 1 oz. of butter.
- 1 whole egg and 1 yolk.
- 2 oz. of Parmesan cheese.
- A little cayenne.
- Pepper and salt.

Method. — Cream the butter well in a basin.

Beat in the eggs, and add the grated cheese.

Season with pepper, salt, and cayenne.

Divide the pastry into two portions, and roll them out as thinly as possible.

Lay one piece on a greased baking-sheet.

Spread it over with the cheese mixture, and lay the other on the top.

148 Mark it with the back of a knife in strips, one inch wide and three inches long.

Brush over with beaten egg, and bake in a quick oven, until the paste is cooked. Cut out the strips with a sharp knife.

Dish them on a folded napkin, and sprinkle them with grated cheese.

Cheese Straws.

- *Ingredients* — 2 oz. of flour.
- 2 oz. of butter.
- 2 oz. of grated Parmesan cheese.
- The yolk of an egg.
- A little cayenne.
- Pepper and salt.

Method. — Rub the butter lightly into the flour.

Add the grated cheese and seasoning, and mix with the yolk of egg.

If necessary, add another yolk, but no water.

Roll out and cut into fingers about a quarter of an inch wide and two inches long.

Lay them on a greased baking-sheet.

Stamp out with a cutter, the size of an egg-cup, some rounds, and make them into rings by stamping out the middles with a smaller cutter.

Bake the rings and straws a pale fawn colour, and serve them with a bundle of straws placed in each ring.

Gooseberry Turnovers.

- *Ingredients*—Some gooseberries.
- Sugar.
- Short crust.

Method.—Pick off the heads and tails of the gooseberries.

Roll out the paste and cut into rather large rounds.

Wet the edges and put some gooseberries in the middle of each round, with a teaspoonful of sugar.

149 Fold the paste over and press the edges together.

Decorate the edges with a fork or spoon.

Put the turnovers on a greased baking-sheet, and bake in a quick oven for fifteen minutes.

Petit Choux.

- *Ingredients*—5 oz. of flour.
- 2 oz. of butter.
- 3 oz. of castor sugar.
- 3 whole eggs.
- ½ pint of water.

Method.—Rub the flour through a sieve.

Put the butter and water on to boil.

When boiling, stir in the flour and sugar.

Beat well over the fire, until the mixture leaves the sides of the saucepan, then remove the saucepan from the fire and beat in three eggs.

Shape like eggs, with two dessertspoons and a knife dipped in hot water.

Lay the pastry on a greased baking-sheet, and bake in a moderate oven for one hour.

To serve, open the cakes at the side and insert a little whipped cream or preserve.

Decorate by brushing them over with white of egg, or a syrup of sugar and water, and sprinkle with chopped *pistachio* kernels, grated cocoa-nut, or hundreds and thousands.

Apple Turnovers.

Make like gooseberry turnovers, substituting minced apple for gooseberries.

Apple Dumplings.

- *Ingredients*—1 dozen apples.
- 1 lb. of short crust.
- A little moist sugar.

150 *Method.*—Pare the apples and remove the cores; fill the holes with sugar.

Take pieces of paste large enough to cover the apples. Do not roll them, but draw the paste over the apples.

Wet the edges to make them join.

Place the dumplings on a greased tin and bake for about three-quarters of an hour or one hour. The length of time will depend on the kind of apples used.

151

PUDDINGS.

A pudding which is to be boiled should be placed in a well-greased basin, or mould, which it should quite fill. A scalded and floured cloth should be tied securely over it. Some puddings, such as suet, plum, &c., may be cooked without the basin, the mixture being firmly tied in a well-scalded and floured cloth, a little room being allowed for the pudding to swell. When cooked in this way, it is well to put a plate in the saucepan to prevent the pudding sticking to the bottom and burning.

To cook a boiled pudding successfully, the water should be kept briskly boiling during the whole of the time it is cooking, and there should be sufficient water in the saucepan to well cover it. A kettle of boiling water should be at hand to fill up the saucepan as required. In steaming puddings, unless a steamer is used, the water should not be allowed to come more than halfway up the pudding-mould, and must only gently simmer, until the pudding is cooked. The mould used need not be covered with a cloth, but a piece of greased paper should be placed over it to prevent the condensed steam dropping on the pudding. Some puddings require to be steamed very carefully, such as contain custard, for example. A custard pudding will be honeycombed (i.e. full of holes), if the water is allowed to boil; the heat of boiling will curdle the eggs.

Most baked puddings require a moderate oven, particularly such as rice, tapioca, &c.

In preparing suet for puddings, remove the skin, slice the suet, and then chop it finely, using a little flour to 152 prevent it sticking to the knife. Currants must be well washed and dried. Sultanas should be rubbed in flour, and the stalks picked off.

Beef-steak Pudding.

- *Ingredients*—1 lb. of flour.
- ¾ or ½ lb. of suet.
- 1½ lb. of beef or rump steak.

- ½ lb. of bullock's kidney.
- Seasoning.

Method.—Chop the suet finely, mix well with the flour, adding a pinch of salt.

Mix to a paste with cold water.

Roll it out, and line a greased quart-basin, reserving one-third for the cover.

Cut the steak into thin strips, and the kidnéy into slices.

Mix some pepper and salt on a plate, and season the meat nicely.

Roll each piece of meat round a tiny piece of the fat, and place the rolls and the pieces of kidney in the basin.

Pour in rather more than a quarter of a pint of water.

Roll out the remaining piece of paste.

Wet the edges of that in the basin, lay the cover on, and trim round neatly.

Tie over a well-scalded and floured cloth, and boil for four hours.

Oysters are sometimes put in these puddings; they should be bearded, and the hard white part removed.

A rabbit or veal pudding may be made in the same manner. To these add a quarter of a pound of lean ham or bacon.

Where economy must be studied, less suet may be used in making the crust.

153

Suet Pudding.

- *Ingredients*—1 lb. of flour.
- 4, 6, or 8 oz. of finely-chopped suet.
- A pinch of salt, or, if liked, a teaspoonful of baking powder.

Method. — Mix the flour and suet lightly together.

Add the salt.

Mix to a stiff paste with cold water.

Then boil in a well-scalded and floured cloth for three hours.

Sultana Pudding.

- *Ingredients* — 1 lb. of flour.
- ½ lb. of finely-chopped suet.
- ¼ lb. of sultanas.
- ¼ lb. of castor sugar; or, three ounces of moist sugar.
- 2 oz. of candied peel.
- The grated rind of a lemon.
- A pinch of salt.
- 1 egg.
- A little milk.

Method. — Rub the sultanas in flour and pick off the stalks.

Cut the candied peel in small pieces.

Put all the dry ingredients into a basin, and mix with the egg, well beaten, and a little milk.

Boil in a basin or cloth three hours.

Compote of Rice.

- *Ingredients* — ¼ lb. of rice.
- ¼ lb. of sugar.
- 1 pint or more of milk.
- Vanilla or other flavouring.

Method.—Boil the rice in the milk, with the sugar, for 154 twenty minutes; if very stiff, add a little more milk or cream.

Flavour with vanilla, and put into a buttered mould with a well in the centre.

Any fruit may be put in the middle, when it is served.

If oranges are used, boil 1½ gill of water with ¼ lb. of lump sugar, until it sticks to a knife like an icicle.

Peel the oranges, and roll them in it.

If apples are used, boil them gently in one pint of water, with ¼ lb. of sugar.

When tender, add a little cochineal.

Take the apples out, and reduce the syrup to less than a quarter of a pint.

Roll the apples in it.

Queen Victoria Pudding.

- *Ingredients*—¼ lb. of butter.
- ¼ lb. of castor sugar.
- ¼ lb. of flour.
- 2 oz. of chopped peel.
- 1 oz. of blanched and chopped almonds.
- 2 tablespoonfuls of brandy.
- 3 eggs.

Method.—Put the butter and sugar in a basin.

Cream them well together with a wooden spoon.

Add the yolks of the eggs one by one; then the flour, peel, almonds, and brandy.

Beat the whites of the eggs stiffly, and mix them in lightly.

Put the mixture in a well-buttered mould.

Cover with buttered paper, and steam for three hours.

Rice Bars.

- *Ingredients* — ¼ lb. of rice.
- 1 pint of milk.
- 3 oz. of castor sugar.
- Yolks 155 of 2 eggs.
- A little lemon essence.
- 1 whole egg.
- Some bread-crumbs.
- Some red jam.

Method. — Boil the rice in the milk, with the sugar, for half an hour, gently stirring occasionally.

Then remove from the fire and, when cool, beat in the two yolks, and add the lemon essence.

Then spread on a flat dish to cool.

When quite cold, cut into bars.

Brush over with the beaten egg, and cover with bread-crumbs.

Fry in hot fat until lightly coloured.

There should be an equal number of bars.

Spread one half of them with jam, and lay the others on the top.

Rice Cakes.

Put the rice mixtures when hot into well-greased tartlet tins.

Make a small hole in the middle and put in a little jam.

Cover with some more of the rice mixture and let them get cold.

Then egg and bread-crumb them, and fry in hot fat.

Orange Pudding.

- *Ingredients*—The rind and juice of 2 oranges.
- 2 oz. of cake-crumbs rubbed through a sieve.
- 2 oz. of castor sugar.
- 3 eggs.
- 1 gill of milk or cream.

Method.—Put the crumbs in a basin, with the sugar.

Add the grated rind of one orange, and the juice of the two.

156 Beat in the yolks of the three eggs, and add the milk or cream.

Whip the white of one egg to a stiff froth, stir in lightly.

Line a pie-dish with a little good pastry; pour the mixture in.

Bake until set, and of a light brown colour.

Welcome-Guest Pudding.

- *Ingredients*—¼ lb. of suet.
- ¼ lb. of sugar.
- ¼ lb. of cake-crumbs, or ratafias, rubbed through a sieve.
- ¼ lb. of bread-crumbs.
- The rind and juice of one lemon.
- 3 eggs, well beaten.

Method.—Put all the dry ingredients into a basin.

Add the lemon rind and juice, and mix with the eggs.

Put into a well-greased mould.

Cover with buttered paper, and steam for two hours.

Crème Frite.

- *Ingredients —* 1 whole egg.
- 1 white.
- 4 yolks.
- 1 gill of cream.
- 1 gill of milk.
- 1 tablespoonful of castor sugar.
- Flavouring to taste.
- 3 oz. of cake-crumbs.

Method. — Cream the yolks and white well together with the castor sugar.

Add cream, milk, and flavouring.

Strain this custard into a greased pudding-basin, and steam *very gently*, until firm.

Let it get quite cold; then turn it out.

157 Cut into slices about one-third of an inch thick.

Stamp into round or fancy shapes.

Egg and cake-crumb them.

Fry in a frying-basket in hot fat.

Serve on a glass dish, and sprinkle with castor sugar.

Gâteau de Cerise.

- *Ingredients —* 1 lb. of cooking cherries.
- ¼ lb. of lump sugar.
- ½ pint of water.
- A few drops of cochineal.
- ¾ of an ounce-packet of gelatine.
- The juice of one lemon.

Method.—Boil the sugar and water; add the lemon and skim well.

Add the cherries (stoned), and stew for a quarter of an hour.

Melt the gelatine in a little water, and add it to the cherries, with enough cochineal to colour brightly.

Pour the mixture into a border mould.

When set, dip it in hot water for a second or two, and turn on to a glass dish.

Serve with whipped cream in the centre.

Jaune Mange.

- *Ingredients*—½ ounce packet of gelatine.
- ½ pint of water.
- ½ pint of white wine.
- Juice of one and a half lemon.
- Rind of half a lemon.
- 3 oz. of castor sugar.
- 4 yolks.

Method.—Soak the gelatine in the water with the lemon rind.

Then put it in a saucepan with all the other ingredients.

158 Stir over the fire until the custard thickens; but, on no account, let it boil.

Then strain into a wetted mould.

Apple Charlotte.

- *Ingredients*—2 lb. of apples.
- ½ lb. of moist sugar.
- Grated rind of a lemon.
- Slices of broad.

- Some clarified butter.

Method.—Peel and core the apples, and stew them with the sugar, lemon rind, and a quarter pint of water, until reduced to half the quantity.

Take a plain round tin, holding about a pint and a half.

Cut a round of stale bread, about one-eighth of an inch thick; dip it in clarified butter, and lay it in the bottom of the mould.

Line the sides with slices of bread, cut about an inch wide, and one-eighth of an inch thick, and also dipped in butter.

Pour the apple mixture into the mould.

Cover with another round of bread dipped in butter; and bake in a moderately quick oven for three quarters of an hour.

For serving, turn it on to a hot dish, and sprinkle castor sugar over it.

Viennoise Pudding.

- *Ingredients*—5 oz. of stale crumb of bread cut into dice.
- 3 oz. of sultanas.
- ¼ lb. of castor sugar.
- 2 oz. of candied peel.
- Grated rind of a lemon.
- 1 wineglass of sherry.
- ½ pint of milk.
- 2 whole eggs.
- 1 oz. of lump sugar.

159 *Method.*—Put the 1 oz. of lump sugar into an old saucepan, and burn it a dark brown.

Pour in the milk, and stir until it is well coloured and the sugar dissolved.

Beat the eggs well, strain the coloured milk on to them, and add the sherry.

Put all the dry ingredients into a basin, and pour the eggs, milk, and sherry over them.

Let the pudding soak for half an hour.

Then put it into a well-greased pint-mould.

Cover with buttered paper, and steam for one hour and a half.

This pudding is to be served with German sauce (*see* Sauces).

Snow Pudding.

- *Ingredients* — ½ pint of milk.
- 1½ oz. of bread-crumbs.
- Grated rind of a lemon.
- 2 tablespoonfuls of caster sugar.
- 3 eggs.
- 2 tablespoonfuls of strawberry or any other jam.
- A little pastry.

Method. — Put the bread-crumbs into a basin.

Boil the milk, and pour it over them.

Mix in the sugar, one whole egg, and two yolks well beaten, and add the lemon rind.

Line a pint pie-dish with a little pastry.

Spread the jam at the bottom and pour the mixture over.

Bake in a moderate oven until set.

Beat the remaining whites to a stiff froth, with a dessertspoonful of castor sugar; and heap it lightly on the top just before serving.

160

German Puffs.

- *Ingredients*—2 eggs.
- Their weight in castor sugar, and ground rice.
- The grated rind of a lemon.

Method.—Beat the eggs well.

Then stir in, gradually, the castor sugar and ground rice, and add the lemon rind.

Partly fill well-buttered cups, or moulds, with the mixture; and bake in a moderate oven for a quarter of an hour, or twenty minutes.

Serve with a wine or sweet sauce (*see* Sauces).

Apple Amber Pudding.

- *Ingredients*—8 apples.
- 2½ oz. of butter.
- 3 oz. of moist sugar.
- Rind and juice of one lemon.
- 3 eggs.
- A little pastry.

Method.—Wash the apples (they need not be peeled or cored) and cut them into small pieces.

Put them into a stewpan with the butter, sugar, lemon rind and juice, and stew until tender.

Then rub through a hair sieve—the sieve keeps back the peel and pips.

Beat the three yolks into the mixture, and put it into a pint pie-dish lined with a little pastry.

Bake in a moderate oven until set.

Then beat the whites of the eggs to a stiff froth with a dessert-spoonful of castor sugar, and heap on the top.

Put it, again, into a cool oven, until the whites are set.

This pudding may be served either hot or cold.

161

Apple Pudding.

- *Ingredients*—1 lb. of flour.
- 6 or 8 oz. of suet.
- A pinch of salt.
- 1 teaspoonful of baking powder.
- Some apples.
- 3 tablespoonfuls or more of moist sugar.
- The grated rind of a small lemon.
- 2 or 3 cloves.

Method.—Prepare the paste, and line a basin as for beef-steak pudding.

Put in the apples, which should be pared and cored, and sprinkle in the sugar and lemon rind.

Put on the cover of paste, and tie over it a well-scalded and floured cloth.

Boil for one hour, or longer: the length of time will depend on the fruit used.

Any fresh fruit may be substituted for the apple.

Raspberry Pudding.

- *Ingredients*—1 pint of raspberries.
- 3 oz. of sugar.

- Thin slices of bread.
- A little milk.

Method. — Pick the stalks from the raspberries, and mix them with the sugar.

Put them and the bread in alternate layers in a pie-dish, moistening the bread with a little milk.

Bake for half an hour.

Note. — This pudding is very good served with cream or custards. The bottled raspberries may be used instead of fresh fruit.

162

Lemon Pudding.

- *Ingredients* — ½ lb. of bread-crumbs.
- ¼ lb. of finely-chopped suet.
- ¼ lb. of castor sugar.
- The grated rind of one lemon, and the juice of two.
- 2 eggs.
- Enough milk to mix it.

Method. — Put the bread-crumbs and suet into a basin.

Add sugar, grated lemon-rind, and juice.

Mix the pudding with the two eggs, well beaten, and a very little milk.

Boil it for one hour and a half.

This pudding may be served with a wine or sweet sauce (*see* Sauces).

Marmalade Pudding.

- *Ingredients* — ½ lb. of flour.
- ½ lb. of bread-crumbs.
- ½ lb. of finely-chopped suet.
- ½ lb. of moist sugar.
- ½ lb. of marmalade.
- 2 eggs.
- The grated rind of a lemon.

Method. — Put the flour, bread-crumbs, suet, sugar, and lemon rind into a basin.

Mix with the marmalade and two eggs, well beaten, and, if necessary, a little milk.

Put it into a well-greased pudding-basin, and tie over it a scalded and floured cloth.

Boil it for five hours.

General Satisfaction.

- *Ingredients* — 3 sponge cakes.
- 2 tablespoonfuls of strawberry or other jam.
- 1 163 wineglass of sherry.
- Rather more than ½ a pint of milk.
- 4 eggs.
- 1 tablespoonful of sugar.
- A little pastry.

Method. — Line a pie-dish with a little pastry.

Spread the jam at the bottom, and lay on it the sponge cakes, cut in halves.

Beat one whole egg and three yolks well together.

Mix with the sugar and milk, and pour over the sponge cakes.

Bake in a moderate oven until the custard is set.

Beat the three whites stiffly, and lay on the top of the pudding.

Put into a cool oven until the whites are set, and of a pale fawn colour.

This pudding may be served hot or cold.

Marlborough Pudding.

- *Ingredients* — 1 pint of milk.
- 2 tablespoonfuls of flour.
- 2 whole eggs and 2 yolks.
- The grated rind of a lemon.
- 3 oz. of castor sugar.
- 2 oz. of butter.

Method. — Mix the flour smoothly with the milk, and stir over the fire until it boils and thickens.

Add the sugar, the eggs, well beaten, the grated lemon rind, and the butter beaten to a cream.

Line a pie-dish with pastry; pour in the mixture.

Bake in a moderate oven until set.

Yorkshire (or Batter) Pudding.

- *Ingredients* — ½ lb. of flour.
- 1 pint of milk.
- 2 eggs.
- A pinch of salt.

164 *Method.* — Put the flour into a basin, make a hole in the middle, and put in the eggs unbeaten.

Stir smoothly round with a wooden spoon, adding the milk very gradually.

If it is to be served with meat, bake it in a baking-tin, which should be well greased with quite one ounce of butter or clarified dripping.

Curate's Puddings.

- *Ingredients* — The weight of 3 eggs in each sugar, flour, and butter.
- 4 eggs.
- A little flavouring essence of any kind, or the grated rind of a lemon.

Method. — Rub the butter well into the flour.

Add the sugar and the four eggs, well beaten.

Half fill well-buttered cups or moulds, and bake for twenty minutes or half an hour.

Serve with a wine or sweet sauce (*see* Sauces).

Canary Pudding.

- *Ingredients* — 2 tablespoonfuls of flour.
- 2 tablespoonfuls of sugar
- 1 pint of milk.
- 2 eggs.

Method. — Put the milk and sugar on to boil.

Mix the flour with a little cold milk.

When the milk boils pour in the flour, and stir it briskly until it thickens.

When cool, add the two eggs, well beaten.

Bake in a greased pie-dish for half an hour.

Christmas Pudding.

- *Ingredients*—2 lb. of raisins.
- 1 lb. of suet.
- ½ lb. 165 of candied peel.
- ¾ lb. of flour.
- ¼ lb. of bread-crumbs,
- ¼ lb. of moist sugar.
- A little mixed spice.
- Half a nutmeg grated.
- A little lemon rind grated.
- ½ pint of milk.
- 4 eggs.

Method.—Put the dry ingredients into a basin, and mix with the eggs, well beaten, and the milk.

Put into a well-greased basin, and boil ten hours if possible.

Cabinet Pudding.

- *Ingredients*—A few raisins or cherries.
- 1 dozen sponge finger-biscuits.
- 1 oz. of castor sugar.
- 1 pint of milk.
- 2 whole eggs and 2 yolks.
- A little vanilla or other flavouring.

Method.—Decorate a well-buttered pint-and-a-half mould with raisins or preserved cherries.

Beat the eggs and milk well together.

Sweeten with the sugar, and add the flavouring.

Break the cakes into pieces.

Put a quarter of them at the bottom of the mould.

Pour in a little of the custard, then more pieces of cake and more custard, and continue in this way until the mould is full.

Cover with buttered paper, and steam gently for about an hour.

Auntie's Pudding.

- *Ingredients* — ½ lb. of flour.
- ¼ lb. of finely-chopped suet.
- ¼ lb. of currants, well washed and dried.
- 3 oz. 166 of sugar.
- 1 egg.
- A little milk.

Method. — Put all the dry ingredients into a basin.

Mix with the egg, well beaten, and the milk.

Boil in a well-greased basin for an hour and a quarter.

Rhubarb Fool.

- *Ingredients* — 14 sticks of rhubarb.
- ½ lb. of moist sugar (more, if necessary).
- ½ pint of water.
- 1 gill of milk.
- The thin rind of half a lemon.

Method. — Cut the rhubarb in small pieces.

Stew gently with the sugar and water until quite tender.

Rub through a sieve.

Add the milk, and serve cold.

Scrap Pudding.

- *Ingredients*—Some scraps of bread.
- ¼ lb. of moist sugar.
- ¼ lb. of finely-chopped suet.
- The grated rind of a lemon.
- 2 eggs, well beaten.
- ¾ pint of milk.
- Some preserve.

Method.—Dry the bread in a slow oven until it is hard.

Pound it in a mortar, and measure 6 ounces of the powder; mix it with the suet and sugar.

Add the lemon rind; pour over the milk, and add the eggs.

Beat well for a few minutes.

Then put the mixture in layers in a pie-dish alternately with the preserve.

Let the top layer be the pudding mixture.

Bake in a moderate oven until the mixture is set.

167

Bread-and-Cheese Pudding.

- *Ingredients*—6 oz. of dried and powdered bread.
- ½ lb. of grated cheese.
- ½ pint of milk.
- 1 egg, well beaten.
- Pepper and salt.

- A little cayenne.

Method.—Mix all the ingredients together, and bake in a pie-dish until the mixture is set.

Mould of Rice.

- *Ingredients*—½ lb. of rice.
- 1 quart of milk.
- ¼ lb. of moist or castor sugar.

Method.—Boil the rice with the sugar in the milk until it is perfectly soft.

Then put it into a mould.

When cold, turn it out, and serve it with jam.

Norfolk Dumpling.

Ingredients—Some bread dough.

Method.—Make the dough into small round balls.

Drop them into fast-boiling water, and boil quickly for twenty minutes.

Serve immediately, either with meat or with sweet sauce.

Sago Pudding.

- *Ingredients*—1 pint of milk.
- 2 tablespoonfuls of sago.
- 2 tablespoonfuls of sugar.
- 1 egg.

Method. — Simmer the sago in the milk until it thickens.

Add the sugar and the egg, well beaten.

168 Put it into a pie-dish, and bake in a moderate oven for half an hour.

The egg may be omitted if preferred.

Rice Pudding.

- *Ingredients* — 1 pint of milk.
- 2 tablespoonfuls of rice.
- 2 tablespoonfuls of sugar.

Method. — Wash the rice and put it in a pie-dish with the sugar.

Pour the milk over it and let it soak for an hour.

Then bake in a moderate oven for one hour, or more, until the rice is quite cooked.

If eggs are used the rice must be simmered in the milk before they are added, and then poured into the pie-dish.

Tapioca Pudding.

Make like a rice pudding.

Semolina Pudding.

- *Ingredients* — 1 pint of milk.
- 2 tablespoonfuls of semolina.
- 1 tablespoonful of moist sugar.
- An egg, if liked.

Method.—Simmer the semolina in the milk, with the sugar, stirring until it thickens.

Then beat in the egg.

Put in a pie-dish, and bake for half an hour.

Swiss Apple Pudding.

- *Ingredients*—½ lb. of bread-crumbs.
- 3 oz. of suet, finely chopped.
- ¼ lb. of apples, finely minced.
- ¼ lb. of sugar.
- The 169 juice and grated rind of one lemon.
- 1 egg well beaten.

Method.—Mix all the ingredients well together, and bake in a pie-dish for one hour.

Light Sultana Pudding.

- *Ingredients*—3 eggs.
- Their weight in each—butter, flour, and sugar.
- ¼ lb. of sultanas.
- The grated rind of a lemon.

Method.—Beat the butter to a cream.

Mix in gradually the flour and sugar, alternately with the eggs, which should be well beaten.

Then add the sultanas, well cleaned, and the grated lemon rind.

Steam for three hours.

Fun Pudding.

- *Ingredients* — 1 lb. of apples.
- 2 tablespoonfuls of apricot jam.
- 2 tablespoonfuls of sugar.
- 1 oz. of butter.
- 2 dessertspoonfuls of arrowroot.
- 1 pint of milk.

Method. — Peel and core the apples, and slice them very finely.

Lay them at the bottom of a pie-dish, and sprinkle some sugar over them.

Put the butter about them in little pieces, and spread over the apricot jam.

Boil the milk, with the remainder of the sugar, and then stir it into the arrowroot, mixed smoothly with cold milk.

When it thickens, pour over the apricot and apples, and bake for half an hour.

170

Sweet Custard Pudding.

- *Ingredients* — Some apricot jam.
- 3 eggs.
- 1 pint of hot milk.
- 3 tablespoonfuls of castor or moist sugar.
- The grated rind of a lemon.
- A little pastry.

Method. — Line a pie-dish neatly with the pastry, and spread the jam at the bottom.

Beat the eggs with the milk and sugar, and pour over the jam.

Bake in a very moderate oven for about one hour.

Jam Roly-poly Pudding.

- *Ingredients* — 1 lb. of flour.
- 4, 6, or 8 oz. of suet, finely chopped.
- Some red jam.
- 1 teaspoonful of baking powder.

Method. — Put the flour into a basin, and add to it the suet and baking powder.

Mix it with a little cold water and roll it out.

Spread it with the jam, and roll up in the form of a bolster.

Scald and flour a cloth, and sew, or tie, the pudding firmly in it.

Boil for two hours.

Treacle Roly-poly Pudding.

Make like a jam roly-poly, using treacle instead of jam.

Custard Pudding.

- *Ingredients* — 1 pint of hot milk.
- 3 eggs.
- 2 tablespoonfuls of castor sugar.
- A 171 little flavouring essence.
- A little pastry.

Method. — Line a pie-dish with pastry.

Beat the eggs in the milk, with the sugar.

Add the flavouring essence, and strain into the pie-dish.

Bake in a moderate oven for one hour, or until set.

Note.—A richer custard may be made by using five yolks and one whole egg.

Bread-and-Butter Pudding.

- *Ingredients*—Some slices of bread-and-butter.
- 2 tablespoonfuls of sugar.
- 1 pint of milk.
- A few currants, nicely washed.
- 1 or 2 eggs, if liked.

Method.—Put some thin slices of bread-and-butter in the bottom of a pie-dish.

Sprinkle them with sugar and currants.

Lay some more slices on the top, with more sugar and currants.

Pour over the milk, and let it soak for half an hour.

Then bake until set.

If eggs are used, beat them with the milk.

Ginger Pudding.

- *Ingredients*—8 oz. of bread-crumbs.
- 6 oz. of suet, finely chopped.
- ½ lb. of treacle.
- 2 tablespoonfuls of moist sugar.
- 2 teaspoonfuls of ground ginger.
- 2 oz. of flour.
- 1 teaspoonful of baking powder.

Method.—Put the bread-crumbs, suet, flour, ginger, and baking powder into a basin.

Mix with the treacle.

Boil in a basin, or cloth, for two hours.

172

Fig Pudding.

- *Ingredients*—½ lb. of bread-crumbs.
- ¼ lb. of suet, finely chopped.
- 3 oz. of brown sugar.
- 2 oz. of flour.
- The grated rind of a lemon.
- 1 egg.
- ½ lb. of figs.
- A little milk.

Method.—Put the bread-crumbs, suet, and sugar, with the figs, cut small, into a basin.

Add the flour and lemon rind, and mix with the egg, well beaten, and a little milk.

Boil in a well-greased basin for two hours.

Rice Balls.

- *Ingredients*—½ lb. of rice.
- 1 quart of milk or water.
- 3 tablespoonfuls of moist sugar.

Method.—Wash the rice well.

Put it with the sugar and milk, or water, into a large saucepan.

Boil gently for about one hour.

Then press into cups, and turn on to a dish.

These may be served with jam, treacle, butter and sugar, or with a sweet sauce.

Little Batter Puddings.

- *Ingredients* — ¼ lb. of flour.
- ½ pint of milk.
- 1 egg.
- Some jam.

Method. — Put the flour into a bowl, and make a well in the middle.

173 Put in the egg, mix smoothly with a wooden spoon, adding the milk by degrees.

Grease some little patty-pans, and half fill them with the batter.

Bake in a quick oven.

When done, dish on a folded napkin, and put a little jam on each.

Ellen's Pudding.

- *Ingredients* — A little pastry.
- 1 oz. of butter.
- 2 oz. of sugar.
- ½ pint of milk.
- The grated rind of a lemon.
- 1 egg well beaten.
- 2 oz. of cake-crumbs.

Method. — Beat the butter to a cream in a basin.

Mix in the sugar thoroughly.

Add the milk gradually.

Then add the egg and cake-crumbs, and pour the mixture into a pie-dish lined with a little pastry.

It is an improvement to put some jam at the bottom of the dish.

Bake for about half an hour.

Bread-and-Fruit Pudding.

- *Ingredients*—Slices of stale bread.
- 1 pint of raspberries.
- ½ pint of currants.
- ¼ lb. of sugar.

Method.—Line a cake-tin, or pie-dish, with stale bread, cut to fit it nicely.

Stew the fruit with the sugar until nicely cooked.

Pour into the mould, and cover with slices of bread.

Cover it with a plate, with a weight on it, and let it stand until the next day.

174 Turn it out and serve plain, or with custard, whipped cream, or milk thickened with cornflour (*see* Cheap Custard).

Ground-Rice Pudding.

- *Ingredients*—2 tablespoonfuls of ground rice.
- 1 pint of milk.
- 2 oz. of sugar.
- 1 or 2 eggs (these may be omitted if liked).
- A little grated lemon rind, or flavouring essence.

Method.—Boil the milk with the sugar.

Mix the rice smoothly with a little cold milk.

Pour it into the boiling milk, and stir until it thickens.

Add the eggs, well beaten, and the flavouring.

Pour into a pie-dish, and bake for about thirty minutes.

Cold Tapioca Pudding.

- *Ingredients*—5 tablespoonfuls of tapioca.
- 1 quart of milk.
- 4 tablespoonfuls of sugar.
- Lemon, or some other flavouring.

Method.—Soak the tapioca all night in cold water.

The next day pour away the water, and put it, with the milk, into a large stewpan with the sugar.

Simmer gently for one hour.

Then pour it into a wetted basin, or mould.

When set, turn it out, and serve with stewed fruit, jam, or treacle.

Tapioca and Apples.

- *Ingredients*—1 quart of water or milk.
- 4 tablespoonfuls of tapioca.
- 4 tablespoonfuls of sugar.
- 1 lb. of apples.
- The grated rind of a lemon.

175 *Method.*—Soak the tapioca in cold water.

Then simmer it in the milk and water, with the sugar, for thirty minutes.

Add the apples, peeled, cored, and sliced.

Put the mixture into a pie-dish and bake for about one hour in a moderate oven.

Steamed Rice Pudding.

- *Ingredients* — 1 oz. of whole rice.
- 1 tablespoonful of sugar.
- 1 egg.
- ½ pint of milk.

Method. — Wash the rice well, and put it into a saucepan of cold water.

Bring it to the boil, and then pour off the water.

Pour in the milk, and add the sugar.

Simmer until the rice is quite soft.

Remove it from the fire, and when cooled a little, stir in the yolk of the egg.

Beat the white to a stiff froth, and stir it in lightly.

Put the mixture into a well-greased pudding-mould, and steam for thirty minutes.

Ratafia Pudding.

- *Ingredients* — 1 pint of milk.
- 3 eggs.
- 4 sponge cakes.
- ½ lb. of ratafias.

Method. — Boil the milk, and when it has cooled a little add to it the three eggs, well beaten.

Break the sponge cakes and ratafias in pieces, and pour the custard over them.

Decorate a greased mould with raisins, and pour the mixture into the mould.

Cover with greased paper, and steam for two hours.

Serve with sweet or wine sauce.

176

Macaroni Pudding.

- *Ingredients* — ½ lb. of macaroni.
- ¼ lb. of sugar.
- 1 or 2 eggs.
- 1 quart of milk.

Method. — Break the macaroni into pieces and put them into a saucepan of boiling water.

Boil for twenty minutes, and then strain off the water.

Pour in the milk; add the sugar, and simmer gently for ten minutes.

Beat up the eggs and stir them in.

Put the mixture into a buttered pie-dish and bake for about thirty minutes.

Eastern Pudding.

- *Ingredients* — 1 lb. of figs (cut in small dice).
- ¼ lb. of suet.
- ½ lb. of bread-crumbs.
- 2 eggs.
- The grated rind of a lemon.

- 1 wineglass of brandy.
- 3 oz. of sugar.

Method. — Put the figs, suet, bread-crumbs, and grated lemon rind into a basin.

Mix it with the eggs, well beaten, and the brandy, adding a little milk if necessary.

Boil in a greased basin for two hours.

Ground-Barley Pudding.

- *Ingredients* — 1 tablespoonful of ground barley.
- ½ pint of milk.
- 1 tablespoonful of moist sugar.
- 1 egg.

Method. — Mix the barley smoothly with the milk.

177 Put it into a saucepan with the sugar, and bring to the boil, stirring all the time.

Then let it simmer for fifteen minutes.

Remove from the fire, and beat in the yolk of the egg.

Whip the white up stiffly, and stir in lightly.

Pour the mixture into a buttered pie-dish, and bake for fifteen minutes.

Steamed Semolina Pudding.

- *Ingredients* — 3 oz. of semolina.
- 1 pint of milk.
- 2 eggs.
- 2 oz. of moist sugar.

- A little flavouring essence.

Method. — Boil the semolina in the milk, with the sugar, until quite soft.

Then add the flavouring essence and the yolks of the two eggs.

Beat the whites up stiffly and mix them in lightly.

Pour the mixture into a greased pudding-mould, and steam for one hour.

Albert Puddings.

- *Ingredients* — 4 oz. of flour.
- 4 oz. of butter.
- 4 oz. of castor sugar.
- 2 eggs.
- A few drops of vanilla flavouring.

Method. — Work the butter to a cream in a basin, and beat in the flour, sugar, and eggs smoothly.

Add the flavouring essence.

Put the mixture into well-greased cups and bake for about half an hour.

Serve with sweet sauce.

178

Pearl-Barley Pudding.

- *Ingredients* — 1 oz. of pearl barley.
- 1 pint of milk.
- 2 oz. of moist sugar.

Method. — Put the barley to soak in cold water all night.

Then pour away the water and put the barley into a pie-dish.

Add the sugar and milk; and bake in a moderate oven for three hours.

Baked Lemon Pudding.

- *Ingredients* — 1 pint of milk.
- 3 oz. of bread-crumbs.
- 1 egg.
- 3 oz. of moist sugar.
- The juice of a lemon and half the rind, grated.

Method. — Put the crumbs into a basin.

Boil the milk with the butter and sugar, and pour it over the crumbs.

Stir in the egg, well beaten; add the lemon rind and juice.

Pour it into a greased pie-dish, and bake in a moderate oven until set.

West-of-England Pudding.

- *Ingredients* — 3 tablespoonfuls of sago.
- 6 small apples.
- 1 quart of milk.
- 3 oz. of moist sugar.

Method. — Soak the sago in cold water for an hour.

Then simmer it in the milk, with the sugar, for twenty minutes.

Peel and core the apples.

179 Place them in a buttered pie-dish, and pour the sago over them.

Bake in a moderate oven for about one hour.

Pancakes.

- *Ingredients* — ½ lb. of flour.
- 2 eggs.
- 1 pint of milk.
- Some lard, or dripping, for frying.

Method. — Put the flour into a basin, add to it a pinch of salt.

Make a well in the middle and put the two eggs into it; mix them smoothly with the flour; and add the milk very gradually.

Melt the lard, or dripping.

Well season a small frying-pan, about the size of a cheese plate.

Put into it a teaspoonful of the melted fat, and let it run well over the pan.

Then pour in enough batter to cover the pan thinly, and fry it brown, shaking the pan occasionally to keep it from burning.

Then toss it on to the other side; and, when that is fried, turn it on to kitchen paper.

Sprinkle with sugar and lemon juice and roll it up.

Keep it hot while the remainder of the batter is fried in the same way.

If the maker cannot toss the pancakes well, they may be turned with a broad-bladed knife. If they are fried in a larger pan, more fat must be used.

Railway Pudding.

- *Ingredients* — ¼ lb. of flour.
- 2 oz. of castor sugar.
- 2 eggs.
- ½ 180 pint of milk.
- 2 teaspoonfuls of baking powder.

Method. — Mix the flour, sugar, and baking powder in a basin.

Beat the eggs well with the milk, and mix the pudding with them.

Pour into a well-greased Yorkshire-pudding tin; and bake for about thirty minutes.

When done, turn out and cut into squares.

Dish in a circle, with a little jam, or treacle, on each.

Poor Knight's Pudding.

- *Ingredients* — Some small square slices of stale bread.
- Castor sugar.

Method. — Fry the bread in hot fat (*see* French Frying).

Drain on kitchen paper.

Dish in the form of a wreath, the one leaning on the other, and put a little jam on each.

Gooseberry Fool.

- *Ingredients* — 1 quart of gooseberries.
- ¾ lb. of moist sugar.
- ½ pint of water.
- 1 pint of milk or cream.

Method.—Take the tops and stalks from the gooseberries, and boil them with the sugar and water until soft.

Rub them through a hair sieve.

Mix in the milk, or cream, gradually; and serve on a glass dish.

Apricot Pudding.

- *Ingredients*—¼ lb. of finely-chopped suet.
- ½ lb. of bread-crumbs.
- 3 eggs.
- 8 tablespoonfuls of apricot jam.
- 1 181 glass of sherry.
- 2 oz. of sugar.

Method.—Put the suet, bread-crumbs, and sugar into a basin, and mix with the eggs, well beaten, apricot and sherry.

Put the mixture into a greased pudding-mould and boil for two hours.

Stale-Bread Pudding.

- *Ingredients*—½ lb. scraps of bread.
- 1 quart of boiling milk.
- 2 eggs.
- 2 oz. of sugar.
- ¼ lb. of currants.

Method.—Soak the bread in cold water until soft.

Squeeze it quite dry, and beat up with a fork.

Pour the boiling milk over.

Stir in the sugar and eggs, well beaten.

Then stir in the currants.

Bake in a pie-dish for two hours.

Baked Plum Pudding.

- *Ingredients*—¼ lb. of finely-chopped suet.
- ¾ lb. of flour.
- ¼ lb. of raisins, stoned and chopped.
- ¼ lb. of currants.
- 2 oz. of candied peel.
- 2 oz. of moist sugar.
- 1 egg.
- 2 teaspoonfuls of baking powder.
- 1 gill, or more, of milk.

Method.—Put all the dry ingredients into a basin, and mix with the egg and milk; it must be quite stiff.

Bake in a greased baking-tin for one hour.

For serving, cut into squares, and dust them over with castor sugar.

182

Treacle Pudding.

- *Ingredients*—1 lb. of flour.
- ¼ lb of finely-chopped suet.
- ¼ lb. of treacle.
- ½ oz. of ground ginger.
- 1 egg.
- 2 oz. of moist sugar.
- 1½ gill of milk.
- 1 teaspoonful of baking powder.

Method.—Put the dry ingredients into a basin.

Mix with the treacle and the egg well beaten with the milk.

Boil in a greased basin for four hours.

The egg may be omitted, if liked.

Plum Pudding.

- *Ingredients*—¼ lb. of finely-chopped suet.
- ¼ lb. of currants.
- ¼ lb. of raisins, stoned and chopped.
- 6 oz. of flour.
- 6 oz. of bread-crumbs.
- 2 oz. of candied peel.
- 3 oz. of sugar.
- 1 gill of milk.
- 2 eggs.
- ½ teaspoonful of baking powder.

Method.—Put the dry ingredients into a basin, and mix with the eggs and milk, well beaten together.

Boil in a cloth or basin for four hours.

Windsor Pudding.

- *Ingredients*—2 oz. of semolina.
- 1 oz. of candied peel.
- ½ pint of milk.
- ¼ lb. of treacle.

183 *Method.*—Mix the milk smoothly with the semolina.

Then put it into a saucepan and stir until it thickens.

Add the treacle and candied peel; pour it into a pie-dish.

Bake for about thirty minutes.

Spring Pudding.

- *Ingredients* — 1 pint of gooseberries.
- ½ pint of milk.
- 4 oz. of moist sugar.
- Slices of bread-and-butter.

Method. — Stew the gooseberries with a very little water and the sugar for ten minutes.

Dip the bread into the milk, and lay a slice at the bottom of a pie-dish.

Put a layer of gooseberries on it.

Then another slice of bread-and-butter and more gooseberries.

Continue in this manner until the dish is full.

Bake gently for one hour.

Gingerbread Pudding.

- *Ingredients* — ½ lb. of flour.
- ½ lb. of treacle.
- ¼ lb. of finely chopped suet.
- 3 teaspoonfuls of ground ginger.
- ½ teaspoonful of baking powder.
- 2 oz. of candied peel.
- 1 egg.
- A little milk.

Method. — Put the dry ingredients into a basin.

Mix with the egg, well beaten, treacle and milk.

Boil in a greased basin for three hours.

184

Economical Bread Pudding.

- *Ingredients*—½ lb. of scraps of bread.
- ¼ lb. of finely-chopped suet.
- ¼ lb. of currants.
- 3 oz. of moist sugar.
- 1 egg.

Method.—Soak the bread in cold water until soft; squeeze it quite dry.

Beat it up with a fork.

Add to it the suet, sugar, and currants, which should be well washed and dried.

Mix with the egg, well beaten.

Boil in a greased basin for an hour.

Economical Ginger Pudding.

- *Ingredients*—½ lb. of scraps of bread.
- ¼ lb. of finely-chopped suet.
- 2 oz. of moist sugar.
- 2 tablespoonfuls of treacle.
- 3 teaspoonfuls of ground ginger.

Method.—Soak the bread in cold water until quite soft.

Squeeze it dry, and beat with a fork until quite fine.

Add the suet, sugar, and ginger, and mix with the treacle.

Boil in a greased basin for an hour.

Economical Fig Pudding.

- *Ingredients* — ½ lb. of scraps of bread.
- ¼ lb. of finely-chopped suet.
- ½ lb. of figs.
- 1 egg.
- 3 oz. of moist sugar.

Method. — Soak the bread in cold water until quite soft.

Squeeze it dry.

185 Add to it the suet, sugar, and figs, chopped small, and mix with beaten egg.

Boil in a greased basin for one hour.

Economical Lemon Pudding.

Make like preceding recipe, substituting the grated rind and juice of two lemons for the figs.

Currant Pudding.

- *Ingredients* — 3 eggs.
- The same weight of sugar, flour, and bread-crumbs.
- Suet, currants, minced apples.
- A little grated lemon rind.
- A little milk.

Method. — Chop the suet finely, and add to it the sugar, flour, bread-crumbs, minced apple, currants, and grated lemon rind.

Mix with the eggs, well beaten, and a little milk.

Boil in a greased basin for three hours.

Plain Cold Cabinet Pudding.

- *Ingredients*—1 tablespoonful of flour.
- 1½ tablespoonful of arrowroot.
- 1 wineglass of sherry.
- A few raisins.
- 3 stale sponge cakes.
- 1 pint of milk.
- 2 oz. of sugar.

Method.—Put the milk to boil with the sugar.

When boiling, stir in the flour, mixed with a little cold milk.

When it thickens, add the arrowroot, also mixed smoothly with milk.

Boil for three minutes, stirring all the time.

186 Then add to it the sherry.

Cut the raisins in two and stone them.

Decorate a plain round tin with them.

Break up the cakes and put some pieces in the tin.

Pour in some of the thickened milk, then some more pieces of cake, and more milk.

Continue in this way until the mould is full.

Set it aside until quite cold.

Then turn it out, and serve with jam.

Cornflour Pudding.

- *Ingredients* — 2 tablespoonfuls of cornflour.
- 1 pint of milk.
- 2 tablespoonfuls of castor sugar.
- 1 egg, if liked.

Method. — Put the milk on to boil.

Put the cornflour into a pie-dish with the sugar.

Mix smoothly with a little cold milk.

Pour on it the boiling milk, stirring quickly until it thickens.

Add the egg, well beaten, and a little flavouring essence.

Bake in a pie-dish for about thirty minutes.

Swiss Pudding.

- *Ingredients* — 2 lb. of apples.
- ½ lb. of bread-crumbs.
- 3 oz. of moist sugar.
- A little grated lemon rind.
- 1 oz. of butter.

Method. — Peel, core, and slice the apples.

Put a layer of them into a buttered pie-dish.

Sprinkle them with crumbs, lemon rind, and a little sugar, and put small pieces of butter about them.

Put some pieces of apple on the top; sprinkle them also with crumbs, lemon rind, sugar, and butter.

187 Continue in the same way until the dish is full.

Bake until the pudding is nicely browned.

For serving, it may be turned out of the dish.

Brown-Bread Pudding.

- *Ingredients*—1 loaf of brown bread.
- 1 gill of double cream.
- The rind of 1 lemon.
- 3 oz. of castor sugar.
- 1 gill of milk.
- 4 eggs.
- A few drops of essence of vanilla.

Method.—Remove the crust from the loaf, and rub the crumb through a wire sieve.

Put five ounces of the crumbs into a basin with the sugar and grated lemon rind.

Boil the milk, pour it over the crumbs, and add the vanilla essence.

Whip the cream to a stiff froth, and mix it with the pudding, adding also the yolks of the eggs.

Beat the whites of two eggs to a stiff froth, and stir them in lightly.

Put the mixture into a well-greased mould, and steam for an hour and a half.

Diplomatic Pudding.

- *Ingredients*—¼ pint of sweet jelly.
- 1 pint of milk.
- ½ oz. of gelatine.
- 2 sponge cakes.
- 2 oz. of ratafias.
- 1 whole egg, and 4 yolks.

- 2 oz. of sugar.
- A little flavouring essence.

Method. — Soak the gelatine in a little milk.

188 Break the sponge cakes and ratafias, and put them into a basin.

Boil the milk with the sugar.

Beat the eggs, and pour the milk on them.

Strain it into a jug, and put it to stand in a saucepan of boiling water, and stir until the custard coats the spoon.

Then melt the gelatine, add it to the custard, and pour it at once over the cakes.

While the mixture cools, pour a little jelly, coloured with cochineal, into a plain round tin.

When it is set, place a jam-pot, or a smaller tin, on it, and pour some jelly round the sides.

When it is quite firm, pour some boiling water into the jam-pot, or tin, and remove it quickly.

When the custard and cakes are cold, but not set, add the essence, and pour into the mould.

When quite firm, dip the tin in hot water for a second or two, and turn it on to a glass dish.

Pease Pudding.

- *Ingredients* — 1 pint of split peas.
- Pepper and salt.

Method. — Soak the peas overnight.

Tie them in a bag or cloth, leaving room for them to swell.

Cook them with the meat with which they are to be served.

Then drain them in a colander.

Mash them with pepper and salt, and press them into a shape in a vegetable-dish.

Hominy Porridge.

- *Ingredients*—1 pint of milk or water.
- 3 tablespoonfuls of flaked hominy.

Method.—Mix the hominy smoothly with the milk or water.

Stir and cook over the fire for ten minutes.

189

Hominy Pudding.

- *Ingredients*—3 tablespoonfuls of flaked hominy.
- 1 pint of milk.
- 2 tablespoonfuls of sugar.

Method.—Mix the hominy with a little cold milk, and make the remainder boil.

Then stir in the hominy and cook until it thickens.

Add the sugar, pour into a greased pie-dish, and bake for about half an hour.

If liked, one or two eggs may be added to the pudding, with a little flavouring essence.

Note.—The *flaked* hominy is the best for general purposes, as the *granulated* takes many hours boiling before it is properly cooked.

190

VEGETABLES.

The rules for cooking vegetables are very simple, and easily remembered. All vegetables, with the exception of old potatoes, are put into boiling water. Green vegetables must be boiled with the lid off the saucepan, as the steam would discolour them, and the water must *boil, not simmer*. Salt is added, in the proportion of one tablespoonful to every two quarts of water. If the water is very hard, it may be necessary to add a little piece of soda. The lime in hard water discolours green vegetables, and the use of soda is to throw this down. Do not, however, use soda, unless obliged, as too much of it will destroy, to some extent, the flavour of the vegetables. Peas must be boiled gently, as rapid boiling would break their skins. Haricot beans must be boiled gently, for the same reason. Root vegetables take longer to cook than fresh ones. Old potatoes must be put into warm water, as they require gradual cooking, and must be boiled gently, until tender. With that exception, all the others must be put into boiling water. Carrots, turnips, and parsnips are generally cooked with the meat with which they are served, as their flavour is thereby improved.

To Boil Potatoes.

If boiled in their skins, scrub them perfectly clean, and put them into a saucepan with sufficient warm water to cover them.

Sprinkle them with salt and boil them gently for half an hour or more, until very *nearly* tender, but not quite.

191 Then pour the water away.

Peel the potatoes, replace them in the saucepan, sprinkle salt upon them, cover them with a cloth, and put the lid on the saucepan.

Let them stand by the side of the fire to finish cooking in their own steam.

Care must be taken that the potatoes cooked in this way are free from disease. One tainted potato would destroy the flavour of the others.

If cooked without their skins, pare them thinly and treat them in the same manner, pouring off the water when they are very nearly tender, and finish cooking them in their own steam.

If the potatoes are good and are cooked according to these directions, they will be perfectly dry and flowery.

To Steam Potatoes.

Put the potatoes into the steamer, and sprinkle them with salt.

Keep the water in the saucepan underneath quickly boiling the whole time the potatoes are cooking.

If the potatoes are cooked in their skins,[*] peel them when very nearly tender, and put them back in the steamer to finish cooking.

Steaming is one of the simplest and best ways of cooking potatoes. If the potatoes are good and the water is kept briskly boiling, this method cannot fail to be successful.

To Cook New Potatoes.

Put the potatoes into boiling water with some salt, and boil gently for twenty minutes or more, according to their age.

When very nearly tender pour off the water, cover them with a cloth, and set the saucepan by the side of the fire, and finish cooking in their own steam.

192

Baked Potatoes.

Choose nice potatoes, not too large, and scrub them perfectly clean.

Bake them in a moderate oven for about an hour.

Brussels Sprouts.

Trim them nicely and put them in boiling water, adding salt in the proportion of a tablespoonful to every two quarts of water.

Put in a little sugar, or, if the water is hard, a little piece of soda the size of a pea.

Boil them quickly, with the lid off the saucepan, from ten to twenty minutes, according to the size and age of the sprouts.

When tender, drain them quite dry in a colander.

Dry the saucepan and put them back with a little butter, pepper, and salt.

Shake them over the fire for a minute or so, and then serve on a hot dish.

To Boil a Cauliflower.

Soak it in salt and water to draw out any insects, and trim off the outside leaves.

Put it, with the flower downwards, into a saucepan of boiling water with salt in it, and cook from twenty to thirty minutes, according to its age.

Drain it on a sieve or colander.

If liked, it may be served with white or French sauce poured over it (*see* Sauces.)

Green Peas.

Put them into plenty of boiling water, with a little sugar and a sprig or two of mint.

193 Boil gently with the lid off the saucepan for twenty minutes or more, according to their size and age.

Drain them in a colander.

Then put them into a saucepan with a little piece of butter, a tea-spoonful of castor sugar, pepper and salt, and shake them over the fire for a minute or two.

French Beans.

Remove the strings and cut the beans into slices.

Put them into plenty of boiling water, with salt in the proportion of one tablespoonful to every two quarts of water, a little sugar, or, if the water is hard, a small piece of soda about the size of a pea.

Boil quickly for fifteen minutes or longer, according to their age.

Drain in a colander.

Then put them into a saucepan with a small piece of butter, pepper and salt, and shake them over the fire for a minute or two.

Spinach.

Pull off the stalks and wash the spinach well in several waters to remove all grit.

Put it into a saucepan without any water but that which adheres to the leaves, and sprinkle a little salt over it.

Cook with the lid off the saucepan until quite tender, stirring it occasionally.

Drain it in a colander, and wring it dry in a cloth.

Then chop it, or rub it through a wire sieve. The latter method is preferable.

To dress it, mix it in a saucepan over the fire with a little butter, pepper, and salt; a little cream may be used also, care being taken not to make the spinach too moist to serve.

Press it into shape, as a mound or pyramid, in a vegetable dish, and garnish with fried *croutons* of bread.

Asparagus.

Cut the asparagus all the same length, and scrape the white part lightly.

Tie it together and put it in boiling water, to which salt has been added, in the proportion of one tablespoonful to two quarts of water.

Add also half an ounce of butter.

Boil gently with the lid off the saucepan for half an hour, until the green part is tender — very young asparagus will not take so long.

Dish on toast; if liked, French or white sauce may be poured over the green ends.

Jerusalem Artichokes.

Peel them, and throw them into boiling water, with salt in the proportion of one tablespoonful to every two quarts of water.

Boil gently with the lid on the saucepan for about fifteen or twenty minutes, until quite tender.

They may be served plain, or with French or white sauce poured over them.

They should be sent to table quickly, or they will be discoloured.

Carrots.

Scrape them and put them into boiling water with salt in it, in the proportion of one tablespoonful to every two quarts of water.

Boil gently with the lid on the saucepan until they are quite tender.

New carrots will take about twenty minutes, old ones an hour or more, according to their age and size.

When they are served with boiled meat, they are generally cooked with it. New carrots are sometimes boiled in second stock.

195 When tender, they are put on a hot vegetable dish, the stock is rapidly boiled down to a glaze, and poured over them.

Turnips.

Boil according to directions given for cooking carrots. Turnips generally take about half an hour; but the time depends on their age and size. If liked, they may be rubbed through a wire sieve, and mashed with butter, pepper, and salt.

Parsnips.

Cook like carrots. They may be served plain, or rubbed through a wire sieve and mashed with butter, pepper, and salt.

Haricot Beans.

Soak them overnight.

Put them into boiling water with a small piece of butter and a small onion.

Boil gently from three to four hours until quite tender.

Drain them, and before serving shake them over the fire with a little butter, pepper, and salt.

Spanish Onions.

First blanch them by putting them into cold water and bringing it to the boil.

Then throw away the water.

Rinse the onions, sprinkle some salt over them, and put them into fresh water.

Boil gently from two to three hours, until perfectly tender.

Drain them, and serve, if liked, with French, Italian or white sauce.

Spanish onions are sometimes boiled in stock, or milk which is afterwards used to make the sauce.

196

Celery.

Clean the celery thoroughly, and tie it in bundles.

Put it in boiling water, milk, or stock, with a little salt and butter, and simmer gently for twenty minutes or more, until quite tender.

Dish on a piece of toast.

If liked, a sauce may be made with the liquor in which the celery has been cooked, and poured over it.

Vegetable Marrows.

Peel the marrows thinly, and cut them in quarters, removing the seeds.

Put them in boiling water, with salt in the proportion of one tablespoonful to every two quarts of water, and boil gently until tender.

They may be served, if desired, with French or white sauce poured over them.

Marrows are very nice when boiled in milk; the milk can afterwards be used to make the sauce.

Cabbage.

Take off the outer decayed leaves, and soak the cabbage in salt and water, to draw out any insects. If very large, cut into quarters.

Put into boiling water, to which salt should be added, in the proportion of a tablespoonful to every two quarts of water. If the water is hard, a piece of soda the size of a bean should be added.

Boil quickly—with the lid off the saucepan—for half an hour, or more, until tender.

Drain well in a colander before serving.

197

Broad Beans.

Put them, when shelled, into boiling water, to which salt should be added in the proportion of a tablespoonful to every two quarts of water.

Boil gently, from fifteen minutes to half an hour, according to their size and age.

When tender, pour the water away, and shake them in the saucepan over the fire, with a little butter or dripping, pepper, and salt.

Tomatoes.

These are better baked than boiled: boiling destroys their flavour.

Put them on a baking-tin, greased with butter or dripping.

Sprinkle over them a little pepper and salt, and cover them with a greased paper.

Put them in a moderate oven, for about ten minutes or a quarter of an hour.

Seakale.

Tie it in bundles, and put into boiling water, with a little butter, and also some salt, in the proportion of a tablespoonful to every two quarts of water.

Boil, with the lid off the saucepan, until the seakale is tender.

Drain, and serve on toast. French or white sauce may be poured over it.

Seakale is sometimes boiled in milk, which should afterwards be used to make the sauce.

Mushrooms.

Peel the mushrooms; rinse them to remove any grit, and cut off the ends of the stalks.

198 Put them on a greased baking-tin, with the stalks upwards, and put some little bits of butter on each mushroom, with a little pepper and salt.

Cover them with buttered paper, and bake them in a moderate oven from ten to twenty minutes, until tender.

Serve on a hot dish, with the gravy poured over them.

Stewed Mushrooms.

Peel and rinse the mushrooms, and cut off the ends of the stalks.

Stew them gently in water, stock, or milk, until quite tender, adding pepper and salt to taste.

Then thicken the gravy with a little flour, and let it cook well, stirring carefully.

Before serving, stir in a little cream or butter.

Fried Potatoes.

Take thin peelings of potatoes, and twist into fancy shapes, or cut the potatoes into thin slices.

Dry them well in a cloth, and drop them into hot fat (*see* French Frying) until quite crisp, and of a light brown colour.

Remove them with a fish-slice or colander-spoon, and drain them on kitchen paper.

Tomato Farni.

- *Ingredients*—6 or 8 ripe tomatoes.
- 1 oz. of butter.
- ½ oz. of flour.
- 1 gill of stock or milk.
- 1 dessertspoonful of chopped parsley.
- 1 dessertspoonful of chopped cooked ham.
- 1 dessertspoonful of grated Parmesan cheese.
- A few button mushrooms, chopped.
- A 199 few drops of lemon juice.
- Some white and browned bread-crumbs.
- Pepper and salt.

Method.—Melt the butter in a small stewpan.

Mix in the flour smoothly.

Then add the stock or milk; stir and cook well.

Then mix in sufficient white bread-crumbs to make the mixture stiff.

Add the parsley, mushrooms, cheese, ham, lemon-juice, pepper, and salt.

Scoop out the top of each tomato.

Pile a little of the stuffing on each, and sprinkle a few browned bread-crumbs over.

Put them on a greased baking-sheet, and cook them in a moderate oven for about a quarter of an hour.

Cauliflower au gratin.

- *Ingredients*—1 cauliflower.
- 1 oz. of butter.
- 1 oz. of flour.
- 1 gill of water.
- 2 tablespoonfuls of cream.
- 2 oz. of grated Parmesan cheese.
- Pepper, salt, and a little cayenne.

Method.—Boil the cauliflower; remove the green leaves.

Place it, with the flower upwards, in a vegetable-dish, and press it into an oval shape.

Melt the butter in a small stewpan.

Mix the flour in smoothly.

Add the water; stir and cook well.

Then add the cream, and one ounce of Parmesan cheese, pepper, salt, and cayenne.

Pour the sauce over the cauliflower.

Sprinkle the remainder of the cheese over it, and brown, either with a salamander or in a quick oven.

200

Potato Croquettes.

- *Ingredients*—2 lb. of potatoes.
- 2 oz. of butter.
- 2 eggs.

- Pepper and salt.
- Some white bread-crumbs.

Method. — Boil the potatoes, and rub them through a wire sieve.

Mash them well with the butter, pepper, and salt.

Mix in one egg, well beaten.

Flour the hands very slightly, and form the mixture in balls, or any other shape preferred.

Brush them over with beaten egg, and cover them with crumbs.

Slightly mould them again when the crumbs are on them.

Fry in a frying-basket, in hot fat (*see* French Frying).

Garnish with fried parsley.

Salsify Patties.

- *Ingredients* — Some patty-cases, made as for oysters.
- ½ lb. of salsify.
- 1 oz. of flour.
- ½ pint of milk.
- 2 tablespoonfuls of cream.
- A few drops of lemon juice.
- Pepper and salt.
- A little cayenne.

Method. — Cook the salsify in milk or water until tender.

Then cut it into small pieces.

Melt the butter in a small stewpan, mix in the flour smoothly.

Then add the milk; stir and cook well.

201 Mix in the cream and let it boil in the sauce.

Then add the lemon juice, seasoning, and salsify.

Fill the patty-cases with the mixture, and put a lid on each.

Tomatoes au gratin.

- *Ingredients* — 1½ lb. of tomatoes.
- 1 pint of bread-crumbs.
- 2 oz. of butter.
- Pepper and salt to taste.

Method. — Slice the tomatoes, and put a layer of them in the bottom of a pie-dish.

Cover them with crumbs; sprinkle with pepper and salt, and place small pieces of butter on them.

Then put another layer of tomatoes, covering them in the same way with crumbs.

Use up all the tomatoes and crumbs in this way, letting the last layer be of crumbs.

Bake in a quick oven for about twenty minutes.

Mashed Potatoes.

- *Ingredients* — 1 oz. of butter to every pound of potatoes.
- 1 tablespoonful of cream, if possible.
- Pepper and salt to taste.

Method. — The potatoes should be well cooked, and be dry and floury.

Put them quickly through a wire sieve.

Mix them well in a saucepan with the butter, cream, and seasoning.

Make them quite hot.

Heap them in a mound-like form in a vegetable dish, and smooth over with a knife.

Mashed Potatoes (a plainer way).

Add to the potatoes, while in the saucepan, some butter or dripping.

202 Season with pepper and salt.

Beat with a fork until perfectly smooth and free from lumps.

Where economy must be studied, nice beef dripping will be found an excellent substitute for butter.

Potato Balls.

Form some mashed potatoes into balls.

Brush them over with beaten egg.

Put them on a greased baking-tin, and bake in a quick oven until brown.

Serve garnished with parsley.

This is a nice way of using up cold potatoes.

Flaked Potatoes.

Rub some nicely-cooked floury potatoes through a wire sieve into a hot vegetable dish. This must be done quickly, that the potatoes may be served quite hot.

Rice for a Curry.

Well wash some Patna rice. Throw it into plenty of quickly-boiling water with salt in it, and boil until the rice is nearly cooked, but not quite. This will take from eight to ten minutes. Strain the rice on a sieve and pour hot water over it, rinsing it well. Then put it in the saucepan again, cover it and let it stand in a hot place to finish cooking in its own steam.

203

SOUPS.

These are very valuable preparations, and are useful to the poor as well as to the rich, as many of the most nutritious soups are the cheapest. Pea soup, haricot soup, and lentil soup are all rich in nourishment, and may be made at a trifling cost, stock not being *necessary* for their manufacture. The boilings from meat, when not too salt, may be used with advantage in making these soups; but if this is not available, they may be made quite well with water; and, if carefully prepared, will have all the flavour of a meat soup.

In making stock for meat soups, it must be borne in mind that in order to extract the juices from the meat it must be put into *cold* water, which should be heated very gradually, and only allowed to *simmer*. In this way a rich stock is procured, as all the virtue of the meat is drawn into the water. Boiling would produce a poor and flavourless stock, as the extreme heat applied, by hardening the albumen, would tend to keep in the juices of the meat instead of drawing them out.

In making stock from bones, the method to be pursued is quite the opposite. Bones must be boiled, otherwise the gelatine in them will not be extracted; simmering would be of little use. The gelatine can only be thoroughly extracted when they are boiled at higher pressure than is possible in ordinary cookery. Bones contain so much gelatine that after they have been once used in stock they should be broken up in pieces and again boiled, so that the gelatine from the *inside* may also be extracted.

204 An economical cook will often make excellent stock for soup from bones alone, with the addition of suitable vegetables for flavouring.

First Stock for Clear Soup.

- *Ingredients*—4 lb. of shin of beef, or 2 lb. of shin of beef and 2 lb. of knuckle of veal.
- 5 pints of water.

- 2 carrots.
- 2 turnips.
- 1 onion.
- The white part of a leek.
- 1 dozen peppercorns.
- 1 sprig of parsley, thyme, and marjoram.
- A bay leaf.
- Pepper and salt.

Method. — Cut the meat into pieces about one inch in size.

Break up the bone and remove the marrow.

Put bones and meat into a stockpot with the cold water.

Let them soak for half an hour.

Then put the pot on the fire; add some salt and pepper to it, and gently simmer the contents for half an hour.

Next put in the vegetables sliced, and the herbs tied together.

Simmer for 4½ hours longer, skimming occasionally.

Strain into a clean pan, and set aside to get cold.

White Stock.

This may be made by the directions in the preceding recipe, using white meat instead of beef; knuckle of veal is considered the stock meat for white soup. Knuckle of veal and a rabbit make excellent stock.

Very good economical white stock may be made by using bones only in making the stock, and no meat; use a ham-bone, if possible, with the others, as this gives a nice flavour.

205

Second Stock.

Take any scraps of cooked or uncooked meat; any bones, cooked or uncooked, to make second stock. Allow one pint of water to every pound of meat and bones, and vegetables in the same proportion as for first stock. The bones should be broken up. Boil gently until all the virtue is extracted from the meat, bones, and vegetables. The contents of the stockpot should be emptied into a pan every night, and the stock strained from the meat, bones, and vegetables. These should be looked over, and the bones, meat, &c., which are of no further use removed; the remainder should be set aside to use with fresh stock material. Bones may be boiled for a very long time before the gelatine will be perfectly extracted.

Second stock, when cold, should be a stiff jelly, in consequence of the gelatine contained in the bones.

White Stock from Bones uncooked.

- *Ingredients* — 4 lb. of uncooked bones, with a ham-bone, if possible, amongst them.
- 5 pints of water.
- 2 carrots.
- 2 turnips.
- 1 large onion.
- Half a head of celery.
- 1 sprig of parsley.
- Thyme, marjoram, and a bay leaf.
- 1 blade of mace.

Method. — Break up the bones and put them with the vegetables, sliced, into a stockpot with the water; boil gently for five hours, adding pepper and salt to taste. Then strain into a clean pan.

206

Clear Soup.

- *Ingredients* — 2 quarts of first stock.
- ¾ lb. of gravy beef.
- The white and shell of one egg.

Method. — Remove *all* the fat from the stock. If it is in a jelly, take off as much as possible with an iron spoon, and remove the remainder by washing the top of the stock with a cloth dipped in very hot water.

Scrape the beef finely and soak it in two tablespoonfuls of cold water to loosen the juices.

Put the stock in a stewpan and add the beef to it, the white and shell of the egg, and a very tiny piece of each kind of vegetable used in making the stock.

Whisk over the fire until the stock begins to simmer.

Then leave off stirring and let it well boil up.

Remove it from the fire and put it on one side for a crust to form.

Tie a clean cloth to the four legs of a chair turned upside down.

Pour some boiling water through it into a basin, to ensure it being perfectly clean.

Then put a clean basin underneath and pour all the contents of the stewpan on to the cloth. The first time the soup runs through it will be cloudy, because the filter made by the beef and egg will not have settled at the bottom of the cloth.

Take the soup away; put a clean basin under the cloth, and pour the soup slowly through.

If this is carefully done the soup will be quite brilliant the second time of straining, and will not require to go through the cloth again.

Julienne Soup.

This is a clear soup with shred vegetables served in it.

Scrape some carrots and take thin parings of them.

Cut these into very thin strips.

207 Take some thin slices of turnip and cut them into strips of the same length.

Boil the turnips for five minutes, and the carrots for fifteen minutes.

Consommée au Royal.

This is clear soup with a savoury custard in it.

Savoury Custard.

- *Ingredients*—1 whole egg.
- 1 yolk.
- ½ gill of clear soup.
- Pepper and salt to taste.

Method.—Beat the eggs and soup together and strain them into a greased gallipot.

Cover them with buttered paper and steam very gently for a quarter of an hour until the custard is firm.

Let it cool, then turn it out. Cut into thin slices.

Stamp into dice or diamonds and serve them in the clear soup.

If the custard is not very gently steamed it will be full of holes, and useless for this purpose.

Consommée à la Princesse.

Serve small *quenelles* (see *Quenelles* of Veal), made in teaspoons, and nicely poached, in the clear soup.

Friar Tuck.

Make one quart of clear soup boiling hot. Beat two eggs well. When the soup is quite boiling, strain the eggs through a pointed strainer into it.

Celery Soup.

- *Ingredients* — 2 quarts of second stock.
- 4 heads of celery.
- 4 oz. 208 of flour.
- ½ pint of cream or good milk.

Method. — Wash the celery well and cut it in pieces.

Simmer it in the stock for half an hour or more until quite tender.

Make a thickening of the flour; pour it into the soup and boil, stirring, for three minutes.

Then rub through a sieve.

Put it into the saucepan again.

Add the cream, stir and let it boil up.

Serve with fried *croutons* of bread.

Oyster Soup.

- *Ingredients* — 2 dozen oysters.
- 1½ pint of white stock.
- 2 oz. of butter.
- 2 oz. of flour.
- 1 teaspoonful of anchovy sauce.
- A few drops of lemon juice.

- Pepper and salt.

Method. — Beard the oysters and cut them in two.

Put the beards into the stock and simmer them in it for a few minutes.

Melt the butter in another stewpan; mix in the flour smoothly; pour in the stock; stir and boil well.

Add the cream and let it boil in the soup.

Strain the oyster liquor and scald the oysters in it.

Put them in the soup and add the anchovy sauce and lemon juice.

Haricot Soup.

- *Ingredients* — 1 pint of haricot beans.
- 1 onion.
- 2 quarts of water.
- 1 pint of milk.
- ½ oz. of dripping.
- Pepper and salt to taste.

209 *Method.* — Soak the beans over night in cold water.

Boil them with the onion, dripping, pepper and salt, in three pints of water, from three to four hours, until quite soft.

Rub them with their liquor through a wire sieve.

Add the milk, and make the soup hot, stirring it over the fire until it boils.

Serve with fried *croutons* of bread.

Note. — This soup is much improved if it is rubbed through a *hair* sieve after it has been through the wire sieve.

Green Pea Purée.

- *Ingredients* — 2 pints of shelled peas.
- A large handful of pea-shells.
- 2 or 3 leaves of spinach.
- 2 or 3 sprigs of parsley.
- A few young onions.
- A sprig of mint.
- A small piece of soda.
- 1 lump of sugar.
- 3 pints of second stock.
- 2 tablespoonfuls of arrowroot.
- Pepper and salt to taste.

Method. — Wash the shells well, and put them, with the peas and other vegetables, into boiling water, to which is added the soda and the sugar.

When quite tender, drain off the water, and put the vegetables into the stock, which should be made boiling.

Let it boil up.

Then rub through a hair sieve.

Make the soup hot; thicken with arrow-root; and, in serving, add, if liked, a little cream, or glaze.

Potage à l'Américaine.

- *Ingredients* — 3 pints of second stock.
- 2 tablespoonfuls of crushed tapioca.
- 8 good-sized tomatoes.

210 *Method.* — Put the stock into a stewpan on the fire to boil.

When boiling, stir in the crushed tapioca.

Stir and cook for about ten minutes, until it is transparent.

Rub the tomatoes through a hair sieve.

Add them to the stock.

Boil for about two minutes and the soup will be ready to serve.

Cream may be added if liked.

Onion Soup.

- *Ingredients*—6 Spanish onions.
- 3 oz. of butter.
- 2 quarts of water.
- 3 oz. of flour.
- ½ pint of milk or cream.
- Pepper and salt to taste.

Method.—Peel the onions, and cut them in slices.

Fry them in the butter, but do not let them discolour.

Then boil them in the water until quite soft.

Rub them with their liquor through a hair sieve.

Put this *purée* into a stewpan on the fire to boil.

When boiling, stir in a thickening of the flour.

Stir and cook well.

Then add the milk or cream, pepper, and salt, to taste.

Let the soup boil up, and it is ready.

Serve with fried *croutons* of bread.

Tapioca Cream.

- *Ingredients*—1½ pint of white stock.
- 1 oz. of crushed tapioca.

- 2 tablespoonfuls of cream.
- The yolks of 2 eggs.
- Pepper and salt to taste.

Method.—Put the stock into a stewpan on the fire to boil.

211 When boiling, sprinkle in the crushed tapioca.

Stir and cook well for about ten minutes.

Beat the yolks lightly with the cream, and strain them.

Let the stock cool a little, and then add two or three tablespoonfuls of it gradually to the eggs and cream.

Pour the eggs and cream into the stock, and stir over the fire until the eggs thicken like custard.

Care must be taken that the stock does not boil after the eggs are in it, as that would curdle them.

Palestine Soup.

- *Ingredients*—3 lb. of Jerusalem artichokes.
- 2 quarts of stock; or the liquor mutton or veal has been boiled in.
- 1 onion.
- 1 turnip.
- ½ head of celery.
- ½ pint of cream, or good milk.
- Pepper and salt to taste.

Method.—Peel and cut the artichokes and other vegetables into slices.

Boil them in the stock until tender.

Rub through a hair sieve.

Add the cream, and boil it in the soup.

Add pepper and salt; and serve with fried *croutons* of bread.

Potato Purée.

- *Ingredients* — 1 lb. of potatoes.
- 1 onion.
- 1 stick of celery.
- 1½ pint of white stock.
- ½ pint of cream or milk.
- Pepper and salt to taste.

Method. — Peel the potatoes, and cut them, as well as the onion and the celery, into thin slices.

212 Put them in the stock, and simmer gently until tender.

Rub through a tammy-cloth or hair sieve.

Add the cream to the soup, and make it hot.

Serve with fried *croutons* of bread.

Egyptian Purée or Lentil Soup.

- *Ingredients* — 1 pint of Egyptian lentils.
- 1 good-sized onion.
- Carrot and turnip.
- 3 sticks of celery, or one dessertspoonful of celery seed tied in a piece of muslin.
- 2 quarts of water, or liquor from a leg of mutton.
- Pepper and salt.

Method. — Wash the lentils thoroughly.

Then boil them in the water with the vegetables, cut in small pieces, from two to three hours, stirring occasionally; when quite cooked, rub through a wire sieve; season to taste.

Make the soup hot in a stewpan, stirring all the time.

Serve with fried *croutons* of bread.

Note. — This soup is much improved if it is rubbed through a hair sieve, after it has been through the wire sieve.

Pea Soup.

Make according to directions given in preceding recipe, substituting split peas for lentils.

Calf-tail Soup.

- *Ingredients* — 4 calves' tails.
- 1 carrot.
- 1 turnip.
- 1 onion.
- 1 sprig of thyme, parsley, and marjoram.
- A little celery.
- 1 213 small clove of garlic.
- 1 dozen peppercorns.
- 4 oz. flour.
- 2 quarts of stock.
- Salt.

Method. — Cut the tails into joints.

Put them into a stewpan, with the water; when it simmers put in the vegetables, &c., and cook very gently for four hours.

Remove the pieces of tail, and let the stock get cold.

Then remove the fat, and thicken the stock with the flour.

Serve with the pieces of tail in it. A wineglass of sherry may be added if liked.

Ox-tail Soup.

- *Ingredients* — 1 ox-tail.
- 3 quarts of stock.
- 1 carrot, 1 turnip, and 1 onion.
- Half a head of celery.
- 1 slice of ham.
- 1 sprig of parsley, thyme, and marjoram.
- 2 bay leaves.
- 3 cloves.
- Pepper and salt.
- 2 oz. of butter.
- 1 wineglass of sherry.
- A few drops of lemon juice.
- 4 oz. of flour.

Method. — Cut the ox-tail into joints.

Fry them in the butter, with the vegetables, cut in pieces.

Put the tail and vegetables into a large saucepan with the stock, pepper, salt, and cloves.

Boil for very nearly four hours.

Then strain the stock.

214 Remove the pieces of tail, and put them on one side.

When the stock is quite cold, remove the fat perfectly and put the stock in a stewpan on the fire to boil.

When it boils, stir in a thickening made of the flour.

Stir, and cook the flour well.

Then add the sherry and lemon juice. Serve with the pieces of tail in it.

Sheep's-head Soup.

- *Ingredients* — 1 sheep's head.
- 3 quarts of water.
- 1 large carrot.
- 1 large turnip.
- 1 large onion.
- 1 sprig of parsley, thyme, and marjoram.
- 1 head of celery.
- 4 oz. of flour.
- 1 teaspoonful of minced parsley.
- 2 tablespoonfuls of bread-crumbs.
- 1 egg.
- Pepper and salt.

Method. — Split open the head, and clean it thoroughly.

Remove the tongue and brains, and blanch the head by putting it into cold water and bringing it to the boil.

Throw the water away, and rinse the head well.

Put it into a large saucepan with the three quarts of water and the vegetables, cut in small pieces.

Boil gently for five hours.

Then strain into a basin, and rub the meat and vegetables through a wire sieve.

When cold, remove the fat from the stock, and add the pulped vegetables and meat.

Make the soup hot, and stir in a thickening made of the flour.

Boil the flour well, stirring all the time.

Serve the soup with forcemeat balls in it.

215 *To make the Forcemeat Balls.* — Boil the tongue and brains separately.

Chop them up; mix them with the bread-crumbs, pepper, salt, and the minced parsley, and bind with the egg.

Make into balls, and roll them in flour; place them on a greased baking-sheet and bake until brown.

Put them in a soup-tureen, and pour the soup over them.

Tapioca Soup.

- *Ingredients* — 2 oz. of tapioca.
- 4 pints of second stock.

Method. — Wash the tapioca well, and throw it into the stock when boiling.

Simmer gently for half an hour, stirring occasionally.

Ox-cheek Soup.

- *Ingredients* — 1 ox-cheek.
- Some cold water, allowing 1 pint to every pound of meat and 1 quart over.
- 2 carrots.
- 2 turnips.
- 2 onions.
- Half a head of celery.
- 1 sprig of parsley, thyme, and marjoram.
- 2 bay leaves.
- Pepper and salt.
- Flour.
- If possible, a ham-bone.
- A few drops of lemon juice.

Method. — Cut up the cheek, and put it with the bone and vegetables into a stockpot to boil gently for five hours, skimming occasionally.

Then strain the stock into a clean pan and set it aside to get cold.

When cold, carefully remove all the fat.

Put the stock into a stewpan on the fire to boil.

216 When boiling, stir in a thickening made of the flour, mixed smoothly with cold water. Use one ounce of flour to every pint of stock.

Stir and boil the flour in the stock for three minutes.

Add to it a squeeze of lemon juice, and serve the soup with neat pieces of the cheek, about one inch in size, in it.

The remains of the cheek may be reboiled, with fresh vegetables, to make a plain second stock.

Giblet Soup.

- *Ingredients*—2 sets of goose or 4 sets of duck giblets.
- ¼ of a head of celery.
- 1 carrot.
- 1 turnip.
- 1 onion.
- 2 cloves.
- 1 blade of mace.
- 1 sprig of parsley, thyme, and marjoram.
- 2 quarts of second stock.
- A few drops of lemon juice.
- Pepper and salt.

Method.—Clean the giblets thoroughly, and cut them in pieces.

Put them into a saucepan, with the vegetables sliced, and the stock, and simmer gently for two hours.

Then take out the best pieces of giblet, trim them neatly, and set them aside.

Simmer the soup for half an hour longer.

Then add to it a thickening of flour, using one ounce of flour to every pint of stock.

Boil and cook the flour well, and add pepper and salt to taste.

Strain the soup into another saucepan.

Add to it the lemon juice, and, if liked, two glasses of Madeira wine; also the pieces of giblet.

Make it quite hot, and it is then ready for serving.

217

Milk Soup.

- *Ingredients*—4 potatoes.
- 2 onions.
- 2 oz. of butter or dripping.
- 3 tablespoonfuls of semolina.
- 1 pint of milk.
- 2 quarts of liquor from meat.
- Pepper and salt to taste.

Method.—Slice the potatoes and onions; add them to the meat liquor, with the butter and pepper and salt to taste, and boil gently for one hour.

Then rub the soup through a wire sieve.

Put it into the saucepan again, and, when boiling, shake into it the semolina and cook for fifteen minutes, stirring occasionally.

When the semolina is cooked the soup is ready.

If this soup is used for children, water may be substituted for the meat liquor if the latter is not available.

Bonne Femme Soup.

- *Ingredients* — 2 lettuces.
- 2 leaves of sorrel.
- 4 sprigs of taragon.
- 2 sprigs of chervil.
- Half a cucumber.
- 2 pints of white stock.
- The yolks of 3 eggs.
- ¼ of a pint of cream.
- The crust of a French roll.
- ½ oz. of butter.

Method. — Wash the lettuce, taragons, and chervil well, and shred them finely.

Peel the cucumber, and shred it also finely.

Melt the butter, and gently *sauté* the vegetables in it for five minutes, taking care they do not discolour.

218 Boil the stock in another saucepan, and, when boiling, pour it on to the vegetables.

Simmer gently until the vegetables are quite tender.

Beat the yolks of the eggs with the cream, and when the stock has cooled a little strain them through a hair sieve into it.

Put the stewpan by the fire, and stir until the eggs thicken, taking care that the stock does not boil, as that would curdle them.

Add pepper and salt to taste, and the soup is ready.

The crust of the French roll should be served in the soup; it should be baked in the oven and then cut into fancy shapes.

Turnip Soup.

- *Ingredients* — 1 quart of stock, or the boilings from mutton.

- 2 lb. of turnips.
- 1 large onion.
- ½ pint of cream, or good milk.
- 2 large slices of bread.
- Pepper and salt to taste.

Method. — Put the bread to soak in a little cold stock.

Pare the turnips and onions, and cut them in pieces.

Boil them gently in the stock, adding (when well soaked) the bread.

When the turnips are cooked, rub the soup through a wire sieve.

Put it again into the saucepan.

Add to it the cream or milk, pepper and salt to taste; and let it well boil up.

Serve with fried or toasted *croutons* of bread.

Rabbit Soup.

- *Ingredients* — 1 large rabbit.
- 2 quarts of water.
- ½ 219 pint of milk or cream.
- 2 good-sized onions.

Method. — Cut the rabbit into joints.

Put them in a stewpan with the onions sliced and the water.

Stew very gently for three hours.

Then strain the stock and remove the fat.

Put it into a clean stewpan and add a thickening of flour, taking one ounce of flour to every pint of soup.

Stir and cook well.

Add the milk or cream and boil it in the soup.

Season with pepper and salt to taste, and serve with fried or toasted bread.

It is an improvement to this soup to cook a ham-bone with the rabbit, or a slice of lean pork.

Hare Soup.

- *Ingredients*—1 hare.
- 1 lb. of gravy beef.
- 1 carrot, 1 turnip, and 1 onion.
- 1 sprig of parsley, thyme, and marjoram.
- 1 bay leaf.
- 1 dozen peppercorns.
- 1 blade of mace and 5 cloves.
- 2 or 3 oz. of butter or dripping.
- 7 pints of water.

Method.—Cut the hare into joints, and the meat into pieces, and fry them in a stewpan in the butter or dripping.

Afterwards fry the vegetables in the same fat.

Then pour in the water, add the mace and pepper-corns, and simmer gently from four to five hours.

Strain the stock and let it get cold.

Remove the fat perfectly, and put it into a clean stewpan on the fire.

When it boils stir in a thickening of flour, using one ounce of flour to every pint of soup.

220 Cook the flour well, and add a little colouring if necessary.

Season to taste, and, just before serving, pour in two glasses of port wine.

Some forcemeat balls should be served in the soup.

Make them with veal forcemeat, mixed with the liver of the hare finely chopped, and bake them in the oven.

Mulligatawny Soup.

- *Ingredients* — 1 rabbit or chicken.
- 2 quarts of second stock.
- 1 onion.
- 1 apple.
- 2 tablespoonfuls of curry powder.
- ½ pint of cream.
- 2 oz. of butter or dripping.
- A few drops of lemon juice.

Method. — Cut the rabbit, or chicken, into joints, and fry them in the butter or dripping.

Remove them when nicely browned, and fry the apple and onion.

Then put the apple, onion, and meat into a stewpan, with the stock, which should be mixed with the curry powder.

Simmer very gently for an hour and a half, until the meat is tender.

Then remove the meat from the stock, and cut it into neat pieces, convenient for serving in the soup, removing all the bone.

Thicken the soup with flour, using about one ounce to every pint of stock.

Boil the flour well in the stock, and then rub the soup through a wire sieve.

Put it into a stewpan, add the cream, and let it boil in the soup.

221 Put in the pieces of meat; and, just before serving, add a squeeze of lemon juice.

Serve nicely boiled rice with this soup (*see* Rice for Curry).

Parsnip Soup.

- *Ingredients* — 2 quarts of stock.
- 2 lb. of parsnips.
- If possible, ½ pint of cream.
- Pepper and salt to taste.

Method. — Slice the parsnips and put them into boiling stock.

Simmer them for one hour, or more, until quite tender.

Then rub the soup through a wire sieve.

Add the cream to it, and pepper and salt to taste.

Put it into a clean stewpan.

Boil up once more and it is ready.

Serve with fried *croutons* of bread.

Red Lentil Soup.

- *Ingredients* — 1 pint of Egyptian lentils.
- 1 large carrot.
- 3 onions.
- 2 lb. of parsnips.
- 1 sprig of parsley.
- 2 or 3 large crusts of bread.
- 2 quarts of water.
- Pepper and salt to taste.

Method. — Wash the lentils well.

Boil with the vegetables, cut in slices, and the bread, for two hours or more; stirring occasionally when the lentils are nearly cooked, as they are apt to stick to the bottom of the saucepan and burn.

Rub the soup through a wire sieve, adding pepper and salt to taste.

Make it hot again, stirring all the time, and it is ready to serve.

222

Mock-Turtle Soup.

- *Ingredients*—Half a calf's head.
- 3 oz. of butter.
- 1 shalot.
- Half-a-dozen mushrooms.
- 1 carrot.
- ½ a head of celery.
- 1 leek.
- 1 onion.
- 1 small turnip.
- 1 sprig of parsley, thyme, and marjoram.
- 1 bay leaf.
- 1 blade of mace.
- 5 cloves.
- 3 oz. of flour.
- 2 wineglasses of sherry.
- 1 dozen forcemeat balls.
- 4 quarts of water.
- ¼ lb. of ham.
- Pepper and salt to taste.

Method.—Wash the calf's head thoroughly.

Cut all the flesh from the bones and tie it in a cloth.

Put it, with the bones and water, into a large saucepan and let it simmer gently, stirring occasionally for three and a half hours.

Then take out the calf's head and strain the stock into a clean pan.

Let it get cold, and then carefully remove all the fat.

Then put the butter into a stewpan, and fry in it the ham and vegetables, cut into slices, with the herbs, mace, cloves, &c.

When they are fried, put in the flour and fry till a light brown, stirring it to keep it from burning.

Then pour in the stock and stir until it boils.

223 Add pepper and salt to taste; put it by the side of the fire to simmer for half an hour.

Remove all scum, or fat, as it rises.

Then strain the stock into another stewpan.

Cut part of the calf's head into neat pieces and add it to the stock.

Pour in the sherry and lemon juice, and add the forcemeat balls.

Let the soup just come to the boil, and it is ready for serving. The forcemeat balls should be made of veal stuffing, and should be either fried or baked.

They should not be too large.

It is better to make this soup the day before it is wanted.

Pot-au-Feu.

- *Ingredients* — 4 lb. of sticking of beef, or 4 lb. of ox cheek without the bone.
- 2 large carrots.
- 1 head of celery.
- 3 onions.
- 2 turnips.
- 3 sprigs of parsley, thyme, and marjoram.
- 3 cloves.
- 6 quarts of water.
- 2 oz. of crushed tapioca, or sago.
- Pepper and salt to taste.

Method. — Tie the meat firmly into shape with string.

Put it into a large saucepan with the water.

When it boils, add a teaspoonful of salt.

Simmer the meat gently for nearly two hours.

Clean the vegetables thoroughly, tying the celery, parsnips, and carrots together.

Add them, with the exception of the cabbage, to the meat, and simmer gently for two hours more.

Then add the cabbage, cleaned and trimmed; it should be cut in two, and tied together with string.

224 Simmer until it is tender, adding pepper and salt to taste.

The meat is then served with the carrots, turnips, and parsnips, as a garnish, and a little of the liquor poured round for gravy.

The cabbage is served in a vegetable dish.

To make the soup, put two quarts of the liquor into a saucepan. When it boils sprinkle in the sago, or tapioca, and cook for fifteen minutes, stirring occasionally.

Dr. Kitchener's Broth.

- *Ingredients*—4 oz. of Scotch barley.
- 4 oz. of sliced onions.
- 2 oz. of dripping.
- 3 oz. bacon.
- 4 oz. oatmeal.
- 5 quarts of the liquor from meat.

Method.—Wash the barley, and soak it in water for two hours.

Put the meat liquor on to boil.

When boiling, add the barley and the onions.

Let it boil gently for an hour and a half.

Then put the dripping into another saucepan, and fry the bacon in it.

Then add, by degrees, the oatmeal; stir until it forms a paste.

Then pour in the broth.

Season with pepper and salt to taste, and set it by the side of the fire to simmer for thirty minutes; the soup is then ready.

Crowdie.

- *Ingredients* — 1 gallon of liquor from meat.
- ¼ pint of oatmeal.
- 1 onion.
- Pepper and salt.

225 *Method.* — Put the liquor into a saucepan on the fire.

Mix the oatmeal to a paste with cold water.

Pour it into the liquor when boiling.

Stir until it thickens.

Add the onion, finely-chopped, and pepper and salt to taste.

Boil gently, stirring occasionally, for half an hour.

226

BREAD AND CAKES.

In making bread be careful that the yeast is good; otherwise the bread may be heavy. The German and French yeast will do quite as well as the brewers', and are generally more easily procured. The French yeast is the closest and strongest, but, though less is required, bread made with it will take longer to rise than that made with German. The yeast may be tested by mixing it with a little sugar; if it is good, it has the power of dissolving the sugar to a syrup. Everything made with yeast should be allowed a proper time to rise. A quartern loaf will generally be ready to make up in about two hours after the dough is set, but the time of rising will vary according to circumstances—for example, in cold weather it may not rise so quickly as in hot. For making bread, warm the pan or tub the dough is to be mixed in, but do not make it hot. Take care that the flour is dry, and free from lumps. The water used must be warmed, but care must be taken that it is neither too hot nor too cold. A certain amount of warmth is necessary for the growth of the yeast, but too great heat kills it. The water, therefore, should be lukewarm. When the dough is mixed, sprinkle the top with a little flour to prevent a crust forming; the pan should then be covered with a cloth and placed on a chair in a warm place, free from draught. It may be placed with advantage before the oven or boiler, but should not be put directly in front of a fire. When the dough is exposed to too great a heat it gets moist and sticky, is very difficult to make up, and is heavy when baked. When the dough has risen 227 sufficiently, it should be well kneaded, and then made up into loaves. These loaves are then set on floured tins to rise in a warm place for about twenty minutes before they are baked. The oven should be very hot for the first twenty minutes, and then very much moderated: a sharp heat is necessary at first to throw up the bread; but the rest of the time the heat applied should be moderate. The same heat is required in baking cakes: a sharp heat at first, to throw them up, and moderate afterwards, so that they may get cooked through without the crust burning. The sugar in cakes causes them to burn very quickly. It is, therefore, a wise precaution to line the tin, even for a plain cake, with foolscap paper.

Currants used in cakes should be well washed and dried before they are used, and any stones removed from them. Sultanas should be rubbed in flour, and the stalks picked off. Raisins should be stoned, and cut in two or three pieces.

To cream butter is to work it about in a basin with the hand, or wooden spoon, until it is the consistency of cream.

The cake tins should be kept in a dry place, and before using should be well greased, especially at the bottom.

A Quartern Loaf.

- *Ingredients* — 3½ lb. of flour.
- 1¾ pint of water.
- ½ oz. of salt.
- 1 oz. of German yeast.

Method. — Put 3 lb. of flour into the pan.

Make the water lukewarm, and mix it with the yeast.

Make a hole in the middle of the flour, and mix in the water smoothly and gradually.

Knead lightly for a minute or two.

Sprinkle with a little flour, and set to rise in a warm place for about two hours.

228 Then knead well for about a quarter of an hour, working in the remainder of the flour.

Make it into a loaf, and put it in or on a floured tin.

Set it to rise for about twenty minutes, and then bake.

The oven must be very hot for the first few minutes, and then the heat must be *much* lowered, that the bread may get well cooked through.

Vienna Bread.

- *Ingredients* — 2 lb. of Vienna flour.
- 2 oz. of butter.
- 1 oz. of German yeast.
- 1 pint of milk.
- 1 teaspoonful of salt.

Method. — Rub the butter well into the flour, and add the salt.

Make the milk tepid, and mix smoothly with the German yeast.

Make a well in the middle of the flour, and stir in the milk smoothly.

Knead very lightly for a minute, and then put the dough to rise in a warm place for two hours.

When it has well risen, make it into rolls or fancy twists.

Set them to rise on floured tins for about ten minutes.

Then bake in a quick oven from ten to twenty minutes, according to their size.

When nearly cooked, brush them with a little milk or white of egg to glaze them.

Unfermented Bread.

- *Ingredients* — 1 lb. of flour.
- 2 heaped teaspoonfuls of baking powder.
- Enough water to mix a dough.

229 *Method.* — Put the flour in a pan.

Add the baking powder and mix up with the water.

Make it into small loaves quickly, and bake in a quick oven for about half an hour.

Milk Rolls.

- *Ingredients* — 1 lb. of flour.
- 2 oz. of butter.
- 2 heaped teaspoonfuls of baking powder.
- Enough milk to mix to a dough.

Method. — Rub the butter into the flour lightly.

Add the baking powder, and mix with the milk.

Make into small rolls as quickly as possible, and bake for a few minutes in a quick oven.

Brush over with a little milk to glaze them.

Pound Cake.

- *Ingredients* — 10 oz. of flour.
- 8 oz. of butter.
- 8 oz. of castor sugar.
- 2 oz. of candied peel.
- ¼ lb. of sultanas.
- 4 large eggs.
- Grated rind of a lemon.

Method. — Rub the flour and sugar through a sieve.

Beat the butter to a cream in a basin.

Mix in a little flour and sugar.

Then a well-beaten egg.

Next more flour, sugar, and another egg.

Continue mixing in the same way until the flour, sugar, and eggs are all well blended together.

Add the other ingredients, and put into a well-greased cake-tin lined with buttered paper.

Bake for about two hours.

230

Queen Cakes.

- *Ingredients* — 6 oz. of flour.
- 4 oz. of butter.
- 4 oz. of sugar.
- 4 eggs.
- A few currants.
- Grated rind of a lemon.

Method. — Cream the butter.

Mix in the flour, sugar, and eggs, according to directions given in preceding recipe.

Add the lemon rind, and partly fill small well-greased Queen-cake tins with the mixture.

Sprinkle a few currants on the top of each.

Bake in a moderately quick oven for about a quarter of an hour.

Rock Cakes.

- *Ingredients* — 1 lb. of flour.
- 6 oz. of butter.
- 6 oz. of castor sugar.
- ½ lb. of currants.
- 2 oz. of candied peel.
- 1 teaspoonful of baking powder.
- 2 eggs.
- Grated rind of a lemon.

Method. — Rub the butter into the flour.

Add the sugar, currants, and other ingredients.

Mix very stiffly with the eggs, well beaten.

Put in rough heaps on a well-greased baking-sheet.

Bake in a quick oven for a quarter of an hour.

Plain Rock Cakes.

- *Ingredients* — 1 lb. of flour.
- ¼ lb. of moist sugar.
- ¼ lb. of currants.
- ¼ lb. 231 of butter, lard, or dripping.
- 1 egg.
- A little milk.

Method. — Rub the butter or dripping into the flour.

Add the other dry ingredients.

Mix stiffly with the egg, well beaten, and a little milk.

Put in little rough heaps on a well-greased baking-tin.

Bake in a quick oven for about a quarter of an hour.

Plain Seed Cake.

- *Ingredients* — 10 oz. of flour.
- 3 oz. of butter or clarified dripping.
- 1 teaspoonful of caraway seeds.
- 3 oz. of castor sugar.
- 2 teaspoonfuls of baking powder.
- 1 egg.
- ¾ gill of milk.
- ½ saltspoonful of salt.

Method.—Rub the fat well into the flour.

Add all the other dry ingredients.

Mix with the egg and milk, well beaten.

Bake in a well-greased cake-tin for about an hour.

Sultana Cake.

- *Ingredients*—10 oz. of flour.
- ¼ lb. butter.
- ¼ lb. of castor sugar.
- ¼ lb. of sultana raisins.
- 1 oz. of candied peel.
- 2 eggs.
- 1 teaspoonful of baking powder.
- ½ gill of milk.
- Grated rind of a lemon.

Method.—Rub the butter well into the flour.

Add all the other dry ingredients.

232 Mix with the milk and yolks of the eggs, well beaten together.

Beat the whites of the eggs to a stiff froth, and mix them in lightly.

Put the mixture in a well-greased cake-tin.

Bake for about one hour and a half.

Plain Plum Cake.

- *Ingredients*—1 lb. of flour.
- ¼ lb. of dripping.
- ¼ lb. of currants.
- ½ pint of milk.
- ¼ lb. of sugar.

- 2 teaspoonfuls of baking powder.

Method. — Rub the dripping into the flour.

Add the other dry ingredients.

Mix with the milk.

Bake in a well-greased cake-tin for about one hour and a quarter.

Rice Cake.

- *Ingredients* — 8 oz. of ground rice.
- 6 oz. of castor sugar.
- 4 eggs.
- Grated rind of a lemon.

Method. — Beat the eggs well with a whisk.

Mix in gradually the castor sugar and rice, and add the lemon rind.

Bake in a well-greased baking-tin in a quick oven for about one hour.

Cornflour Cake.

- *Ingredients* — ¼ lb. of cornflour.
- ¼ lb. of castor sugar.
- 2 oz. of butter.
- 1 teaspoonful of baking powder.
- 2 eggs.

233 *Method.* — Beat the butter to a cream.

Then mix in the sugar.

Add the two eggs, and beat all well together.

Lastly, stir in the cornflour and add the baking powder.

Put the mixture into a well-greased cake-tin, and bake for about three-quarters of an hour.

Scones.

- *Ingredients* — 1 lb. of flour.
- 2 oz. of castor sugar.
- 3 oz. of butter.
- ½ pint of milk.
- 2 teaspoonfuls of baking powder, or ¼ oz. of cream of tartar, and
- 1 teaspoonful of carbonate of soda.

Method. — Rub the butter into the flour.

Add the other dry ingredients.

Mix lightly with the milk.

Divide the dough into two pieces.

Make each piece into a ball.

Roll it out to about three-quarters of an inch in thickness.

Cut into triangular-shaped pieces.

Bake on a greased baking-tin for about twenty minutes.

Brush them over with a little white of egg or milk to glaze them.

Currant Cake.

- *Ingredients* — 10 oz. of flour.
- ¼ lb. of currants.
- ¼ lb. of sugar.
- 3 oz. of butter.
- 1 egg.

- 2 teaspoonfuls of baking powder.
- ¼ pint of milk.
- A little grated lemon rind.

234 *Method.* — Rub the butter into the flour until like fine bread-crumbs.

Add the sugar and currants — the currants should be well washed and dried — also the baking powder and lemon rind.

Mix with the beaten egg and milk.

Bake it at once, in a greased cake-tin lined with paper, for one hour and a half.

Luncheon Cake.

- *Ingredients* — 1 lb. of flour.
- 4 oz. of butter, lard, or dripping.
- ¼ lb. of sultanas.
- ¼ lb. of currants.
- 6 oz. of sugar.
- 2 oz. of candied peel.
- 2 eggs.
- Rather less than a ¼ pint of milk.
- 2 teaspoonfuls of baking powder.
- 1 oz. of lump sugar.
- Grated lemon rind.

Method. — Put the lump sugar in a saucepan and burn it brown.

Pour in the milk and stir until it is coloured.

Then strain it and let it get cold.

Put the flour into a basin.

Rub the butter lightly into it.

Add the sultanas (well cleaned), and the rest of the dry ingredients.

Mix with the eggs well beaten, and the milk.

Put it into a well-greased tin, which should be lined with paper.

Bake from one hour and a half to two hours.

235

Gingerbread.

- *Ingredients*—2 lb. of flour.
- 2 lb. of treacle.
- ½ lb. moist sugar.
- 3 eggs.
- ½ oz. of carbonate of soda.
- 8 oz. of butter.
- 2 oz. of ginger.
- ½ a cup of water.
- 2 oz. of candied peel.

Method.—Put the flour, sugar, ginger, candied peel, and carbonate of soda into a basin.

Warm the treacle, water, and butter in a saucepan.

Mix with the dry ingredients and add the eggs, well beaten.

Partly fill a well-greased Yorkshire-pudding tin.

Smooth over with a knife dipped in hot water, and score with a knife.

Bake in a moderate oven for about an hour and a half.

Sponge Cake.

- *Ingredients*—4 oz. of flour.

- 5 eggs.
- 4 oz. of castor sugar.

Method. — Oil the cake-mould, and dust it over with castor sugar.

Beat the eggs and sugar for about twenty minutes until they rise and are quite light; this may be done over hot water, care being taken that the heat is not too great to cook the eggs.

Dry and sift the flour, and stir it lightly in.

Pour into the mould and bake in a moderate oven for about one hour.

236

Sponge Roll.

- *Ingredients* — 5 eggs.
- The weight of 4 eggs in castor sugar.
- Of 3 in flour.
- Some jam.

Method. — Beat the eggs to a cream.

Add the sugar and then the flour, lightly.

Have a baking-tin ready greased with butter, and lined with greased paper.

Pour in the mixture; spread it over and bake it till a light fawn colour.

Then turn it on to a cloth.

Spread with the jam melted and roll up quickly.

Seed Cake.

- *Ingredients* — 1 lb. of flour.

- 4 oz. of butter.
- 6 oz. of castor sugar.
- ½ oz. of caraway seeds.
- ¼ pint of milk.
- 1 teaspoonful of baking powder.
- 2 eggs.

Method.—Rub the butter into the flour.

Add the castor sugar and seeds.

Mix with the yolks and milk beaten together.

Beat the whites stiffly and stir in lightly.

Bake in a nicely prepared tin for about one hour and a half.

Madeira Cake.

- *Ingredients*—10 oz. of flour.
- 10 oz. of lump sugar.
- ¼ lb. of butter.
- 6 eggs.
- ½ gill of water.

237 *Method.*—Boil the water and sugar to a syrup.

Pour when hot, but not boiling, on to the eggs and beat over hot water until light.

Melt the butter and stir it in very lightly with the flour.

Oil a mould and dust it with castor sugar.

Pour in the mixture, and bake from one hour and a half to two hours.

Buns.

- *Ingredients*—16 oz. of flour.
- ½ oz. of yeast.
- ½ pint of milk.
- 2 oz. of sugar.
- 2 oz. of sultanas.
- 2 oz. of butter.
- 1 egg.

Method.—Put ten ounces of the flour into a basin.

Mix the yeast smoothly with the milk, which should be made tepid.

Stir into the flour.

Beat for five minutes, and set to rise in a warm place for about two hours.

Then beat in the remainder of the flour, sultanas, sugar, butter, and the egg.

Set to rise for about two hours more.

Then form into buns.

Place them on a floured tin, and let them rise for ten minutes.

Bake in a very quick oven for about five minutes until nicely coloured.

Boil half an ounce of sugar with half a gill of water, and brush the buns over with this to glaze them.

Dough Cake.

- *Ingredients*—½ quartern of dough.
- ¼ lb. of currants.
- ¼ lb. 238 of moist sugar.
- ¼ lb. of clarified dripping.

Method.—Put the dough into a basin.

Beat in the dripping, sugar, and currants.

These should be well washed and dried.

Place in a greased tin, and set to rise for one hour.

Bake in a moderate oven for two hours.

Candied-peel Drops.

- *Ingredients*—½ lb. of flour.
- 3 oz. of butter.
- 3 oz. of candied peel.
- Grated rind of a lemon.
- 1 egg.
- A little milk.
- One teaspoonful of baking powder.
- 3 oz. of sugar.

Method.—Rub the butter into the flour.

Add the sugar, grated lemon rind, baking powder, and the candied peel chopped small.

Mix with the egg, well beaten, and the milk.

Put it in little heaps on a greased baking-tin.

Bake in a quick oven for about fifteen minutes.

Shrewsbury Cakes.

- *Ingredients*—¼ lb. of butter.
- ¼ lb. of castor sugar.
- 6 oz. of flour.
- 1 egg.

Method. — Cream the butter and sugar.

Add to them the egg, well beaten.

Then stir in the flour.

Knead it to a dough.

Roll out, and cut into small round cakes with a cutter.

Place them on a greased baking-sheet.

Bake in a moderate oven from fifteen to twenty minutes.

239

Oatmeal Biscuits.

- *Ingredients* — 7 oz. of flour.
- 3 oz. of oatmeal.
- 3 oz. of castor sugar.
- 3 oz. of lard, dripping, or butter.
- ¼ teaspoonful of baking powder.
- 1 egg.
- 1 tablespoonful of water.

Method. — Put the flour, oatmeal, sugar, and baking-powder into a basin.

Mix them with the fat melted, and the egg beaten with the water.

Knead lightly into a dough.

Roll it out, and cut into round cakes.

Place them on a greased baking-tin.

Bake in a moderate oven for about twenty minutes.

Shortbread.

- *Ingredients* — ¼ lb. of flour.

- 1 oz. of castor sugar.
- 2 oz. of butter.

Method. — Put the flour and sugar into a basin.

Melt the butter, and mix them with it.

Knead lightly.

Roll out, cut the paste into cakes with a knife, and bake for half an hour.

Yorkshire Teacakes.

- *Ingredients* — ¾ lb. of flour.
- 1½ gill of milk.
- 1 oz. of butter.
- 1 egg.
- ½ oz. of German yeast.

240 *Method.* — Put the flour into a basin, and rub the butter into it.

Make the milk tepid, and blend it with the yeast.

Strain it into the flour.

Add the egg.

Beat all well together for a few minutes.

Knead lightly.

Then divide the dough in two.

Make each part into a ball, and put them in floured cake-tins.

Put the cakes in a warm place to rise for one hour, and then bake them for about twenty minutes.

Brush them over with a syrup of sugar and water to glaze them.

Ginger Biscuits.

- *Ingredients* — ½ lb. of flour.
- 2 oz. of butter, lard, or dripping.
- ½ oz. of ground ginger.
- 2 oz. of castor sugar.
- 1 egg, and a little milk.
- ½ teaspoonful of baking powder.

Method. — Rub the butter into the flour until it is like fine breadcrumbs.

Add the sugar and baking powder, and mix with the egg, well beaten, and as much milk as necessary to make it bind.

Roll out, and cut into small round cakes.

Put them on a greased tin.

Bake in a moderate oven for about twenty minutes.

Lemon-rock Cakes.

- *Ingredients* — 1 lb. of flour.
- 3 oz. of butter.
- 5 oz. of castor sugar.
- Grated rind of one lemon and juice of two.
- 1 egg.
- A 241 little milk.
- 1 teaspoonful of baking powder.

Method. — Rub the butter into the flour.

Add the sugar, baking-powder, lemon rind, and juice.

Mix with the egg, well beaten, and as much milk as necessary; the mixture should be very stiff.

Put it in little rough heaps on a greased baking-tin.

Bake in a quick oven for about fifteen minutes.

Soda Cakes.

- *Ingredients*—½ lb. of flour.
- 2 oz. of butter.
- 3 oz. of sugar.
- 1 oz. of candied peel.
- Grated rind of a lemon.
- 1 whole egg.
- If necessary, a little milk.
- ½ a teaspoonful of carbonate soda.

Method.—Rub the butter well into the flour.

Add the sugar, peel, lemon rind, and soda.

Mix with the egg, well beaten, and, if necessary, a little milk; the mixture must be very stiff.

Put it in little rough heaps on a greased baking-tin.

Bake in a quick oven for fifteen minutes.

Gingerbread Cakes.

- *Ingredients*—1 lb. of flour.
- 6 oz. of butter, lard, or dripping.
- 1 oz. of ground ginger.
- 4 oz. of moist sugar.
- ¾ lb. of treacle.

Method.—Put the sugar, treacle, and fat into a saucepan, and melt them.

Put the flour and ginger into a basin.

Mix with the other ingredients.

Roll out, and cut into small cakes.

242 Bake on a greased baking-tin, in a slow oven, for ten or fifteen minutes.

Rice Buns.

- *Ingredients*—¼ lb. of ground rice.
- ¼ lb. of castor sugar.
- 2 oz. of butter.
- 1 egg.
- ½ a teaspoonful of baking powder.
- A little flavouring essence.

Method.—Beat the butter to a cream with the sugar.

Then add the eggs, well beaten, and stir in the ground rice.

Partly fill little greased patty-pans with the mixture, and bake in a moderate oven for a quarter of an hour.

Galettes.

- *Ingredients*—1 lb. of Vienna flour.
- 1 lb. of household flour.
- 1 oz. of yeast.
- ½ lb. of butter.
- 6 eggs.
- ½ a pint of milk.
- A little sugar.

Method.—Make the milk tepid.

Then mix it smoothly with the yeast, and stir it into the household flour.

Knead it to a dough.

Rub the butter into the other flour and beat in the eggs well with the sugar.

Then knead both doughs together.

Put them to rise for about two hours.

When nicely risen, make the dough into buns.

Put them on a floured baking-sheet.

Bake in a quick oven for about ten minutes.

When nearly ready, brush over with a little white of egg to glaze them.

243

JELLIES AND CREAMS.

To clear Jellies.

Take a large saucepan, and see that it is perfectly clean. Put into it all the ingredients for the jelly, and the whites and shells of the eggs. The use of the whites of eggs is to clear the jelly; the shells form a filter through which to strain it. Whisk all together over a quick fire until the jelly begins to simmer; then immediately leave off stirring, and let it well boil up. The heat of the boiling jelly hardens the egg, which rises to the surface in the form of a thick scum, bringing all impurities with it. If the stirring were continued during the boiling it would prevent the scum rising properly, and the jelly would not clear.

When the jelly has well boiled up, remove it from the fire and let it stand for a few minutes till a crust is formed.

To strain it, a chair may be turned upside down, and a cloth tied firmly to its four legs. Any cloth, which is clean, and not too closely woven, will answer the purpose. Put a basin under the cloth, and pour some boiling water through it. This will make it hot, and ensure its being perfectly clean. Change the basin for a clean dry one, and pour the whole contents of the saucepan on to the cloth. The first runnings of the jelly will be cloudy, because the filter which the eggs make will not have settled in the cloth. As soon as the jelly runs slowly, and looks clear, put a clean basin under the cloth, and put the first runnings through it again, very gently, that they may not disturb the filter of egg-shells.

Strain the jelly in a warm place, out of draught. Two eggs are considered sufficient to clarify a quart of jelly, but 244 if the eggs are small it is wise to take a third. If there is not sufficient white of egg, the jelly will not clear.

The jelly should be allowed to get nearly cold before it is put into the moulds. If it is put hot into metal moulds it is likely to become cloudy.

To make Creams.

To make a good cream, it is essential that the cream used should be double; that is, a thick cream that will whip up to a stiff froth. Beat it well with a wire whisk until it will stand on the end of it without dropping. This must be done in a cool place, especially in summer time. Cream is liable to curdle, and turn to butter, if beaten in too warm a temperature. The gelatine must be added last of all. It should be stirred in thoroughly, but quickly; it must not be too hot, or too cold, but just lukewarm. If too hot, it destroys the lightness of the cream; if too cold, it does not mix thoroughly. Pour the cream into a mould as soon as the gelatine is mixed with it, as it begins to set directly. To turn a jelly or cream out of its mould, take a basin of hot water, as hot as the hand can bear, draw the mould quickly through it, letting the water quite cover it for a second. Wipe off all the moisture immediately with a dry cloth. Shake the tin gently, to be sure the contents are free. Lay the dish on the open side of the mould, quickly reverse it, and draw the mould carefully away.

Strawberry Cream.

- *Ingredients* — ½ pint of double cream.
- 1 oz. of amber gelatine, or rather less than ½ oz. of the opaque.
- 2 tablespoonfuls of castor sugar.
- Some strawberries.
- ¼ pint of milk.
- A few drops of cochineal.

245 *Method.* —Soak the gelatine in the milk for about twenty minutes or more.

Then dissolve it by stirring it in a saucepan over the fire.

Rub sufficient strawberries through a hair sieve to make a quarter of a pint of *purée.*

Beat up the cream with the sugar.

Then add the *purée* of fruit, and a few drops of cochineal to colour it.

Lastly stir in the melted gelatine.

Pour the cream at once into a wetted mould.

When quite set, dip it for a second or two into very hot water, and turn it on to a glass dish.

Charlotte Russe.

- *Ingredients* — 1 dozen sponge fingers.
- 1 oz. of glace cherries.
- ½ pint of double cream.
- ½ oz. of amber gelatine melted in a little milk, or less than ¼ oz. of the opaque.
- 2 dessertspoonfuls of castor sugar.
- A few drops of essence of vanilla, or other flavouring.

Method. — First put the gelatine to soak in a little milk.

Then cut the cherries in halves, and place them in a circle round the bottom of a plain round tin, with the cut side uppermost.

Divide the sponge fingers, lengthwise, without breaking them, and trim each one at the side, top, and bottom neatly.

Then line the tin with them, placing them on the top of the cherries, with the brown side next the tin; they should be put close together, and the last should serve as a wedge to keep the others in place.

Beat up the cream stiffly with the sugar.

246 Add the vanilla flavouring and the melted gelatine. This must be neither too hot nor too cold.

Stir it thoroughly, but quickly, into the cream, and pour at once into the prepared tin.

When set, dip the bottom of the tin into hot water for a second or two, and turn it carefully on to a glass dish.

Custard Cream.

- *Ingredients* — ½ pint of double cream.
- 3 tablespoonfuls of castor sugar.
- 1 oz. of amber gelatine, or less than ½ of the opaque.
- 1 whole egg.
- 3 yolks.
- ½ pint of hot milk.
- A few drops of vanilla or other essence.

Method. — Put the gelatine to soak in a little milk.

Then beat the eggs lightly and add them to the milk.

Strain into a jug and add the sugar.

Put the jug into a saucepan of boiling water, and stir until the custard coats the spoon; care must be taken that it does not curdle.

While the custard cools beat up the cream stiffly.

Melt the gelatine, and add it to the custard.

Flavour it, and, when sufficiently cooled, mix the custard and cream thoroughly together.

Pour at once into a wetted mould.

Bohemian Cream.

- *Ingredients* — ½ pint of sweet jelly of any kind.
- ½ pint of double cream.

Method. — Beat the cream stiffly.

Mix with it the jelly, which should be melted, but cold.

Pour into a wetted mould.

247

Wine Jelly.

- *Ingredients* — 1 oz. packet of either Nelson's or Swinbourne's gelatine.
- 1 pint of water.
- ½ pint of sherry.
- ¼ to ½ lb. of lump sugar, according to taste.
- The juice of two lemons.
- The rind of one.
- The whites and shells of 2 large eggs.

Method. — Soak the gelatine in the water with the thin rind of a lemon for three quarters of an hour, if possible.

Then add all the other ingredients.

Clarify and strain (*see* To clear Jellies).

When quite cold pour into a wetted mould.

Calf's-foot Stock.

- *Ingredients* — 2 calf's feet.
- 4 pints of water.

Method. — Cut each foot into four pieces.

Blanch them by putting them in cold water and bringing them to the boil.

Throw the water away, and well wash the feet.

Put them into a saucepan, with four pints of water, and boil gently for five hours.

Then strain the stock from the bones, and set it aside until the next day.

The fat must then be carefully removed, or the stock will not clear.

To turn this into Calf's-foot Jelly, add—

- Half a pint of white wine.
- The rind of 2 and the juice of 4 lemons.
- ¾ lb. of lump sugar.
- The whites and shells of 4 eggs.
- Clarify and strain (*see* To clear Jellies).

248

Pineapple Jelly.

- *Ingredients*—1 pineapple.
- 1 oz. packet of gelatine.
- 1 pint of water.
- ¼ lb. of lump sugar.
- The thin rind of 1 lemon, and the juice of 2.
- The whites and shells of 2 large eggs.

Method.—First soak the gelatine in the water.

Cut up the pineapple and bruise it in a mortar.

Add it, and all the other ingredients, to the gelatine.

Then clarify (*see* To clear Jellies).

Note.—The Grated Pineapple, sold in tins, is excellent for jellies or creams.

Aspic Jelly.

- *Ingredients*—1 oz. packet of gelatine.
- 1 pint of good stock.
- ¼ pint of taragon vinegar.
- ¼ pint of sherry.
- A small carrot, turnip, and onion.
- A sprig of parsley, thyme, and marjoram.
- 2 bay leaves.
- 3 cloves.
- 1 dozen peppercorns.
- A piece of celery.
- A blade of mace.
- A clove of garlic.
- 1 shalot.
- The whites and shells of 2 large eggs.
- Salt to taste.

Method.—Soak the gelatine in the stock.

Then add all the other ingredients and clarify (*see* To clear Jellies).

249

Claret Jelly.

- *Ingredients*—1 oz. packet of gelatine.
- ½ pint of water.
- 1 pint of claret.
- ½ lb. of lump sugar.
- A few drops of cochineal.

Method.—Soak the gelatine in the water.

Add the sugar, and stir over the fire until dissolved.

Pour in the wine, and colour with cochineal.

Strain into a wetted mould.

When firm, dip into hot water for a second or two, and turn on to a glass dish.

Note.—This jelly is not clarified. Cake is usually served with claret jelly.

Orange Jelly.

- *Ingredients*—1 dozen oranges.
- 1 lemon.
- 2 pints of water.
- 1 oz. packet and a half of Swinbourne's or Nelson's opaque gelatine (in summer two packets).
- ½ lb. of lump sugar.

Method.—Soak the gelatine in the water with the thin rind of one lemon and three oranges.

Add the sugar; stir over the fire until the gelatine is dissolved.

Add the juice of the twelve oranges.

Let the jelly boil up, and then strain into a wetted mould.

When firm, dip into hot water for a second or two, and turn on to a glass dish.

Note.—This jelly is not clarified.

250

Strawberry Jelly.

- *Ingredients*—1 quart of strawberries.
- ½ lb. of lump sugar.
- Juice of one lemon.
- 1½ oz. of Swinbourne's or Nelson's opaque gelatine.

- ½ pint of cold water.
- ½ pint of boiling water.
- The whites and shells of 2 large eggs.

Method.—Soak the gelatine in the cold water.

Mash the strawberries to a pulp.

Add them to the gelatine with the sugar and lemon juice.

Pour the boiling water over.

Then put all the ingredients into a saucepan.

Add to them the whites and shells of the eggs, and clarify and strain (*see* To clear Jellies).

Pour into a wetted mould, and set in a cool place until firm.

To turn it out, dip the tin into very hot water for a second or two, and turn it carefully on to a glass dish.

Orange Cream.

- *Ingredients*—1 pint of double cream.
- 4 oranges.
- 4 oz. of sugar.
- 1 oz. packet of gelatine.
- 2 whole eggs.
- Yolks of 4 eggs.
- 1 pint of milk.

Method.—Soak the gelatine in a ¼ pint of milk with the thin rind of one orange.

Strain the juice of the oranges into a cup.

Beat the eggs, and yolks of eggs, with the milk.

Strain into a jug, and add the sugar.

251 Put the jug to stand in a saucepan of boiling water, and stir until the custard coats the spoon.

Melt the gelatine and add it to the custard.

Whip up the cream stiffly, and add to it the orange juice.

When the custard is cool, beat it into the cream, and pour at once into a wetted mould.

If liked, it may be put into a border mould, and served with whipped cream in the middle.

Blancmange.

- *Ingredients*—1 oz. packet of Swinbourne's isinglass.
- 1 pint of milk.
- 1 pint of cream.
- 3 or 4 oz. castor sugar.
- Flavouring essence.

Method.—Soak the isinglass in the milk; add the sugar and stir over the fire until both are dissolved.

Then pour in the cream; stir occasionally until cold.

Add the flavouring essence and pour it into a wetted mould.

Note.—A *blancmange* may be made economically by using less cream and more milk, or using milk only. If it is not stirred until cold, the cream and milk will separate.

Vanilla Cream.

Make a thick cream as for Charlotte Russe, and flavour with vanilla.

Gâteau aux Pommes.

- *Ingredients* — 2 lb. apples.
- 3 oz. moist sugar.
- 1 lemon.
- ½ oz. packet of Swinbourne's or Nelson's gelatine.
- ½ pint of water.
- A few drops of cochineal.

252 *Method.* — Soak the gelatine in half the water.

Wash and slice the apples.

Put them in a stewpan with the sugar and thin lemon rind and juice and remainder of the water.

Stew until soft, then rub through a *hair* sieve.

Melt the gelatine; mix it thoroughly with the apples.

Colour with cochineal, and pour the mixture into a wetted mould.

Note. — This sweet looks very nice when it is made in a border mould. It is then served with whipped cream or white of egg in the middle.

Peaches, prunes, or any suitable fruit may be substituted for the apples.

Compote of Peaches.

- *Ingredients* — 10 oz. of sugar.
- 1 pint of water.
- 1 dozen peaches.
- ½ pint of whipped cream.

Method. — Boil the sugar and water for ten minutes.

Pare the peaches and simmer for about twenty minutes.

Remove carefully and place on a glass dish.

Reduce the syrup and pour over them.

When cold, cover with whipped cream.

Almond Bavarian Cream.

- *Ingredients* — 1 pint of double cream.
- ½ lb. of sweet almonds.
- 1 or 2 drops of essence of almonds.
- 4 oz. of castor sugar.
- ¾ of an ounce packet of gelatine.
- 3 eggs.
- ¾ of a pint of milk.

Method. — Soak the gelatine in the milk.

Blanch and pound the almonds, adding a few drops of orange-flower water to keep them from oiling.

253 Beat the eggs and milk lightly together, and strain into a jug.

Add to them the sugar and almonds.

Put the jug into a saucepan of boiling water, and stir until the custard coats the spoon.

Melt the gelatine, and add it to the custard.

Whip the cream to a stiff froth, and drop in the almond essence.

When the custard is cool, stir it into the cream.

Mix them well together, and pour into a wetted mould.

Stone Cream.

- *Ingredients* — 1 pint of double cream.
- 2 wineglasses of sherry.

- Juice of a lemon.
- ¼ lb. of castor sugar.
- 1 gill of milk.
- 1 oz. of Nelson's or Swinbourne's opaque gelatine.
- A little almond flavouring.

Method.—Soak the gelatine in the milk with the sugar.

Beat the cream up stiffly.

Melt the gelatine; add to it the sherry, lemon juice, and flavouring.

Stir it quickly into the beaten cream.

Pour it into a wetted mould.

When set, dip it into very hot water for a second, and turn it carefully on to a glass dish.

Lemon Sponge.

- *Ingredients*—½ oz. packet of gelatine.
- 1 pint of cold water.
- ½ lb. of lump sugar.
- The thin rind and juice of two lemons.
- The whites of 3 eggs.

254 *Method.*—Soak the gelatine in the water with the rind of the lemon for one hour.

Add the sugar and dissolve it over the fire.

Stir and simmer it for a few minutes.

Strain into a basin and add the lemon juice.

When it begins to set, beat in the whites of the eggs, whipped to a very stiff froth.

Whisk until the whole mixture is light and spongy.

Then heap it on a glass dish.

A little of it may be coloured a pale pink with cochineal; and as a decoration, a few pistachio kernels, blanched and chopped, can be sprinkled over the sponge.

Floating Island.

- *Ingredients* — A round sponge-cake.
- 1 pint of custard (*see* Boiled Custard).
- Red jam.
- The whites of two eggs.
- 1 tablespoonful of castor sugar.
- Some chopped pistachio kernels.
- Some hundreds and thousands.

Method. — Cut the cake horizontally in slices.

Spread them with jam.

Place them on each other in the form of the cake, and spread the top with jam.

Put the cake on a glass dish, and pour the custard over.

Whip the whites of the eggs stiffly with the sugar, and heap on the top of the cake.

Decorate with chopped pistachios and hundreds and thousands.

Maraschino Cream.

- *Ingredients* — 3 yolks of eggs.
- 1 white.
- ½ pint of milk.
- ½ pint of whipped double cream.
- 2 255 tablespoonfuls of castor sugar.

- 1 oz. of amber gelatine, or ½ oz. of the opaque, melted in a little milk.
- 1 small glass of maraschino.

Method.—Make the eggs and milk into a custard (*see* Boiled Custard).

Add to it the sugar and melted gelatine.

When it has cooled, mix it with the cream.

Add the maraschino and pour into a wetted mould previously decorated with a little bright fruit.

When set, dip into hot water for a second or two, and turn it on to a glass dish.

Pistachio Cream.

- *Ingredients*—½ pint of whipped double cream.
- ½ oz. of amber gelatine, or less than ¼ oz. of the opaque, melted in a little milk.
- 1 oz. of castor sugar.
- 2 oz. of pistachio kernels blanched.
- A few drops of vanilla.

Method.—Pound the pistachios in a mortar, and rub them through a sieve.

Then mix them with the cream.

Add a few drops of vanilla, the sugar, and, last of all, the melted gelatine.

Pour it into a wetted mould.

When set, dip it into hot water for a second or two, and turn carefully on to a glass dish.

Croquant of Oranges.

- *Ingredients* — 9 or 10 oranges.
- ½ teacupful of melted sweet jelly.
- A few pistachio kernels, blanched and chopped.
- ½ pint of whipped double cream.
- ½ oz. 256 of amber gelatine, or less than ¼ oz. of the opaque, melted in a little milk.
- 2 oz. of castor sugar.

Method. — Peel and divide six oranges into sections, and carefully remove the white skins.

Dip each piece into the jelly, and line a plain round charlotte Russe tin with them.

Place them to form a star in the bottom of the mould, and fill up any spaces with the chopped pistachio kernels.

Add the juice of three oranges to the whipped cream.

Mix in the sugar, and add, last of all, the melted gelatine.

Pour the cream into the tin.

When set, dip the tin in hot water to loosen the pieces of orange, and then turn carefully on to a glass or silver dish.

Chartreuse de Fruit.

- *Ingredients* — Various fruits, such as strawberries, greengages, cherries, peaches, &c.
- Some strawberry or other cream.
- ½ teacupful of sweet jelly, melted.

Method. — Line a plain charlotte Russe mould tastefully with slices of the different fruits, dipping each piece in the melted jelly.

Then pour in a strawberry or any other cream (*see* Strawberry Cream).

When set, dip the mould into very hot water for a second or two to loosen the fruit, and then turn them on to a glass or silver dish.

Strawberry Charlotte.

- *Ingredients*—Some fine ripe strawberries.
- Some pistachio kernels, blanched and chopped.
- ½ teacupful of melted sweet jelly.
- Some strawberry cream.

257 *Method.*—Line a Charlotte Russe mould tastefully with the strawberries cut in half, dipping them in the jelly, and laying them in the tin with the cut side downwards.

Fill the spaces with the pistachios.

When the strawberries are quite firm, pour in some strawberry cream (*see* Strawberry Cream).

When set, dip into very hot water for a second or two to loosen the fruit, and turn on to a glass or silver dish.

Tipsy Cake.

- *Ingredients*—1 large sponge cake.
- 1 wineglass of sherry.
- 1 wineglass of brandy.
- 1 pint custard (*see* Boiled Custard).
- Some blanched almonds.

Method.—Put the cake on a glass dish.

Soak it with the sherry and brandy.

Pour over the custard, and stick blanched almonds well over it.

Trifle.

- *Ingredients* — 1 Savoy cake.
- 1 pint of double cream.
- 1 pint of rich custard (*see* Boiled Custard).
- Some strawberry or other jam.
- 1 wineglass of sherry.
- 1 wineglass of brandy.
- ½ lb. of macaroons.
- 1 oz. of castor sugar.
- 6 sponge cakes.

Method. — Cut the cake into slices an inch thick.

Lay them on the bottom of a glass dish.

Spread them with jam.

Lay the macaroons on them.

Cover them with sponge cakes.

Soak them with the sherry and brandy, and cover with the custard.

258 Whip the cream very stiffly with the sugar.

Drain it on a sieve.

Before serving, heap the whip on the top of the trifle.

Decorate it with chopped pistachios, and hundreds and thousands.

Apple Flummery.

- *Ingredients* — 2 lb. of apples.
- The rind and juice of a small lemon.
- 5 oz. of sugar.
- ¼ pint of water.
- ½ oz. packet of Nelson's or Swinbourn's gelatine.

- ½ pint of cream.
- Cochineal.

Method.—Cut up the apples, and stew them with the sugar, lemon, and water, until tender.

Rub them through a hair sieve.

While the apples are cooking, soak the gelatine in the cream.

Then stir over the fire until the gelatine is quite dissolved.

Add the cream and gelatine to the apple pulp, and beat all well together.

Colour with cochineal, and pour into a wetted mould.

When firm, dip for a second or two into very hot water, and then turn on to a glass dish.

Apple Cream.

- *Ingredients*—2 lb. of apples.
- ¼ lb. of sugar.
- 1 glass of port wine.
- The rind of a lemon.
- ¼ pint of water.
- ½ pint of cream or milk.
- Cochineal.

259 *Method.*—Wash the apples, and cut them into pieces.

Put them into a stewpan with the lemon rind, sugar, wine, and water.

Stew gently until they are quite tender.

Then rub them through a hair sieve, and colour with cochineal.

Boil the cream or milk and add it to the apple pulp.

Beat them thoroughly together, and serve when cold in a glass dish.

Alpine Snow.

- *Ingredients* — 1½ lb. of apples.
- 3 oz. of castor sugar.
- ¼ pint of water.
- The thin rind and juice of half a lemon.
- The whites of 3 eggs.
- Cochineal.

Method. — Wash the apples and cut them in pieces.

Put them in a stewpan with the water, sugar, lemon rind and juice.

Stew gently until quite tender.

Then rub through a hair sieve.

Whip the whites of the eggs.

When the apple pulp is quite cold, add them to it, and beat until the mixture is a stiff froth.

Colour prettily with cochineal, and heap on a glass dish.

Welsh Custard.

- *Ingredients* — 2 lb. of apples.
- The thin rind of a lemon.
- Juice of half a lemon.
- 3 oz. of castor sugar.
- 2 teaspoonfuls of ground ginger.
- 3 whole eggs.
- ¾ pint of water.

Method. — Wash and cut up the apples.

260 Stew them until tender with the sugar, lemon rind and juice, ginger, and water.

Rub them through a hair sieve (there should be about one pint of pulp if the stewing has been very gentle).

Beat the eggs, and strain them into the apple pulp.

Pour the custard into a jug.

Put it to stand in a saucepan of boiling water, and stir until it thickens, taking care that it does not curdle.

Stir occasionally while it is cooling, and serve in custard glasses or on a glass dish.

Cheap Custard.

- *Ingredients* — 1 tablespoonful of cornflour.
- 1 pint of milk.
- The yolks of 2 eggs.
- 2 oz. of castor sugar.
- Vanilla or other flavouring.

Method. — Put the milk and sugar on to boil.

When boiling, stir in the cornflour, which should be mixed very smoothly with a little cold milk.

Boil, stirring all the time, for ten minutes.

Then remove from the fire, and, when it has cooled a little, beat in the yolks of the eggs.

Stir again over the fire to cook the eggs, but take care they do not curdle.

Flavour to taste, and when cold pour into custard glasses.

A cheaper substitute for custard may be made by omitting the eggs.

Arrowroot Custard.

- *Ingredients* — 1 pint of milk.
- 1 tablespoonful of arrowroot.
- 2 oz. of castor sugar.
- The yolks of 2 eggs.
- Vanilla or other flavouring.

261 *Method.* — Boil the milk with the sugar.

When boiling, pour in the arrowroot, mixed very smoothly with a little cold milk.

Stir until it boils and thickens.

Then remove it from the fire, beat in the yolks and stir until they thicken.

Plain Trifle.

- *Ingredients* — A little red jam.
- 5 sponge cakes.
- 1 doz. ratafias.
- 1 pint of milk.
- The white of an egg.
- 3 eggs.
- 1 oz. of castor sugar.

Method. — Boil the milk with the sugar.

Beat the eggs, and stir the milk on to them.

Strain into a jug.

Place the jug in a saucepan of boiling water, and stir until the custard coats the spoon.

Then let it cool, stirring occasionally.

Cut the cakes in halves; spread them with jam; place them on a dish alternately with the ratafias.

Pour the custard over them, and set aside until quite cold. Decorate with the white of egg beaten stiffly.

Boiled Custard.

- *Ingredients*—¾ pint of new milk.
- Yolks of 5 eggs.
- 3 dessertspoonfuls of castor sugar.
- A little flavouring of vanilla, lemon, or almond.

Method.—Boil the milk with the sugar.

Beat the yolks lightly.

Pour the milk (not too hot) on them, stirring all the time.

262 Strain the custard into a jug, which must be placed in a saucepan of boiling water.

Stir until it coats the spoon.

Great care must be taken that the custard does not curdle; it mast be stirred occasionally while cooling.

A cheaper custard may be made by substituting two whole eggs for the five yolks, or one whole egg and two yolks.

263

SOUFFLÉES AND OMELETS.

The best cooks will sometimes fail in making soufflées, as their manufacture requires the very greatest care and attention. It is also necessary to be able to judge to a nicety the time they will take to cook, because, to be eaten in perfection, they should be served directly they are ready. In making a soufflée, be very careful to take *exact* measure of the different ingredients; a little too much flour, or rather too little milk, may make a great difference in the lightness of it. The flour should be the best Vienna.

Another point to be attended to is to whip up the whites of the eggs as stiffly as possible, and to mix them with the other ingredients very lightly. Bear in mind that the object in beating the whites of eggs is to introduce air into the soufflée; and it is the expansion of the air when the soufflée is cooking which makes it light. If the whites are mixed heavily with the other ingredients, the air which has been whipped into them is beaten out again; and consequently they fail to make the soufflée light. When the soufflée is firm in the middle, it is sufficiently cooked, and should be served with the greatest expedition, as it will begin to sink rapidly. An omelet soufflée, left in the oven two or three minutes over time, will be quite spoilt, and become tough and leathery.

Steamed soufflées are turned out of the tins they are cooked in, and served with a sauce poured round them.

Baked soufflées are served in the tins, which are slipped into a hot metal or silver case, or a napkin is folded round them.

264 Plain omelets are quickly made, and quickly spoiled. Some practice is required to make the plain omelet to perfection, as the art consists in folding the omelet just at the right moment, before the eggs used in them are too much set. The omelet should not be firm throughout, like a pancake, but should be moist and succulent in the middle. A very sharp fire is essential, and the omelet should not take more than three minutes in the making.

Steamed Soufflée.

- *Ingredients*—1 oz. of butter.
- 1 oz. of flour.
- ¼ pint of milk.
- 4 eggs.
- 2 dessertspoonfuls of castor sugar.

Method.—Well grease a soufflée-tin with butter.

Fold a half sheet of kitchen paper in three.

Brush it over with melted butter, and fasten it round the top of the tin, letting it come nearly three inches above it.

Melt the butter in a small stewpan.

Mix in the flour smoothly.

Add the milk, and stir and cook well.

Mix in the sugar, and beat in the yolks of three of the eggs, one by one.

Add a little flavouring essence.

Beat the whites of four eggs to a stiff froth, and stir them in lightly.

Put the mixture at once into the tin.

Cover it with buttered paper, and steam carefully for half an hour.

When done, it will be firm in the middle.

Turn it quickly on to a hot dish, and serve at once, with wine sauce poured round it (*see* Sauces).

265

Cheese Fondu.

- *Ingredients*—1 oz. of butter.

- ½ oz. of flour.
- ¼ pint of milk.
- 3 oz. of grated Parmesan cheese.
- 3 eggs.
- A little pepper, salt, and Cayenne.

Method. — Prepare the tin as for a steamed soufflée.

Melt the butter in a small stewpan.

Mix in the flour smoothly, add the milk, and stir and cook well.

Add the seasoning, and beat in the yolks of two of the eggs.

Then mix in the grated cheese.

Beat the whites of the three eggs to a stiff froth, and stir them in lightly.

Put the mixture at once into the tin, and bake for twenty-five or thirty minutes.

When done, it will be firm in the middle.

Serve in the tin, with a napkin folded round it.

Omelet Soufflée.

- *Ingredients* — 2 yolks and 3 whites of eggs.
- 1 dessertspoonful of castor sugar.
- A little flavouring essence.

Method. — Beat the yolks in a basin with the sugar, and add the essence.

Whip up the whites as stiffly as possible, and mix them lightly with the yolks.

Pour the mixture into a well-greased omelet-pan, and put it into a brisk oven for about three minutes, until of a pale-brown colour.

Turn it on to a hot dish.

Fold it over and serve quickly.

266

A Savoury Omelet Soufflée.

May be made by omitting the flavouring essence, and substituting pepper and salt for the sugar. The omelet should then be served with a rich gravy poured round it.

Cheese Ramequins.

Make a mixture as directed for Cheese Fondu. Partly fill little ramequin cases with it, and bake in a quick oven for a few minutes.

Batter for Fritters (Kromesky).

- *Ingredients*—¼ lb. of flour.
- 1 tablespoonful of oiled butter or salad oil.
- 1 gill of tepid water.
- The white of 1 egg, beaten to a stiff froth.
- If for *sweet* fritters, castor sugar to taste.

Method.—Put the flour into a basin.

Make a hole in the middle, and put in the oil.

Stir smoothly, adding the water by degrees.

Beat until quite smooth.

Then add the beaten white, stirring it in lightly.

Apple Fritters.

Pare some nice apples.

Cut them into slices about a quarter of an inch thick, and stamp out the core with a round cutter.

Lay the rings in the batter, and cover them well with it.

Lift them out with a skewer, and drop them into hot fat (*see* French Frying).

When lightly browned on one side, turn them on to the other.

Drain them on kitchen paper.

Dish on a folded napkin, with castor sugar dusted over them.

267

A Small Savoury Omelet.

- *Ingredients* — 2 or 3 eggs.
- 1 dessertspoonful of finely-chopped parsley.
- 1 oz. of butter.
- Pepper and salt.

Method. — Break the eggs into a basin.

Add to them the parsley, pepper, and salt.

Melt the butter in a small omelet-pan.

Beat the eggs very lightly, and pour them into the pan.

Shake and stir the mixture vigorously until it begins to set.

When some of the egg is set and the other still liquid, tilt the pan, and draw the egg quickly to the one side of it.

Leave it there to set for two or three seconds; then tilt the pan again and fold the omelet, quickly drawing it to the other side of the pan.

As soon as the outside is set, turn it on a hot dish and serve immediately.

To make an omelet successfully, a *very* quick fire is necessary; an omelet should not take more than three minutes to cook.

Larger omelets are made by using more eggs and butter and parsley in proportion.

Chopped cooked ham and kidney may be added to a savoury omelet; also mushrooms and shalots. The latter should be finely chopped, and fried in a little butter before they are used. A cheese omelet is made by adding grated Parmesan or other cheese to the mixture.

268

INVALID COOKERY.

Much attention should be paid to this branch of cookery. The recovery of many sick people depends, to a great extent, on their being able to take a proper amount of nourishment. This they will not be likely to do, unless the food is well cooked, and nicely served.

Everything, for an invalid, should be dressed plainly, and *well cooked*. Highly seasoned meat, rich gravies, sauces, puddings, &c., should be avoided. The digestive organs are weakened by illness, and should not be unduly taxed. All meals should be served punctually; carelessness in this respect has often been the cause of great exhaustion. A good nurse ought to watch her patients carefully, and never allow their strength to sink for want of nourishment at the right time.

It is not wise to prepare too large a quantity of anything at one time; an invalid's appetite is generally very variable.

All fat should be carefully removed from beef-tea and broth before they are served. This can be best done when they are cold.

Great care should be taken to make everything look as tempting as possible. The tray-cloths used, glass, silver, &c., should be spotlessly clean, and bright-looking.

Raw-beef Tea.

- *Ingredients* — Equal quantities in weight of beef and
- cold water.

269 *Method.* — Scrape the beef very finely, and remove the fat.

Soak the beef in the water for about half an hour, moving it occasionally with a fork.

When the juices of the meat are drawn into the water, and it has become a deep-red colour, it is ready for use and should be strained.

This tea is better made from rump or beef steak.

Do not make too much at one time. In hot weather two ounces or a quarter of a pound of meat will be quite sufficient.

Be careful that the meat is perfectly sweet and good.

Beef Tea.

- *Ingredients*—1 lb. of rump or beef steak.
- 1½ pint of cold water.

Method.—Cut the steak into small pieces, and put them into a jar with the water; tie a piece of paper over the top.

Put the jar to stand in a saucepan of boiling water for four hours.

Pour the tea from the beef, and remove the fat when cold; salt can be added to taste.

Mutton Broth.

- *Ingredients*—1 lb. of scrag end of neck of mutton.
- 2 pints of water.
- 1 tablespoonful of rice.
- Salt to taste.

Method.—Cut up the mutton, and put it into a saucepan with the water.

Simmer gently for four hours.

Then strain away from the meat, and set on one side to cool.

When quite cold carefully remove the fat, and put the broth into a clean saucepan.

270 Put it on the fire to boil, and, when boiling, throw in the rice, which should have been well washed.

As soon as the rice is cooked, the broth is ready.

Add salt and pepper to taste.

Clear Barley-water.

- *Ingredients* — 2 oz. of pearl barley.
- A little thin lemon peel.
- 1 pint of boiling water.
- Sugar to taste.

Method. — Wash the barley, and put it into a jug with the lemon peel.

Pour the boiling water over it, and add the sugar.

Let it stand until cold, and then strain it.

Thick Barley-water.

- *Ingredients* — 2 oz. of pearl barley.
- 1 quart of water.
- A little thin lemon peel.
- Sugar to taste.

Method. — Wash the barley, and put it into a saucepan With cold water.

Boil for ten minutes.

Then throw the water away, and wash the barley. This is to blanch it. If this were not done the barley water would have a dark-coloured, unpleasant appearance.

Put it into a saucepan, with the quart of water, and boil gently for two hours.

Sweeten to taste, and then strain it.

Rice Water.

- *Ingredients*—2 oz. of rice.
- 3 pints of water.
- 1 inch of cinnamon.
- Sugar to taste.

271 *Method.*—Wash the rice well, and throw it into three pints of boiling water, with the cinnamon.

Boil gently for two hours.

Sweeten to taste, and strain.

Apple Water.

- *Ingredients*—2 large apples.
- A little thin lemon peel.
- 1 pint of boiling water.
- Sugar to taste.

Method.—Peel and cut up the apples.

Put them into a jug with the lemon peel and sugar.

Pour over the boiling water, and cover close until cold; then strain it.

Lemonade.

- *Ingredients*—2 lemons.
- 4 lumps of sugar.
- 1 pint of boiling water.

Method.—Take the yellow part of the lemon peel, cut very thinly, from one of the lemons.

Then remove the skin completely from them both.

Cut them into slices, and remove the pips.

Put the sliced lemon, thin peel, and sugar, into a jug; pour over the boiling water.

Cover, until cold, and then strain.

A Cup of Arrowroot.

- *Ingredients* — ½ pint of milk.
- 1 dessertspoonful of arrowroot.
- Castor sugar.

Method. — Put the milk into a saucepan on the fire to boil.

Mix the arrowroot very smoothly with a little cold milk; when the milk boils pour in the arrowroot, and stir until the milk has thickened.

Add sugar to taste.

For water arrowroot, substitute water for milk.

272

Arrowroot Pudding.

- *Ingredients* — Cup of arrowroot, made as in foregoing recipe.
- 1 or 2 eggs.
- A little vanilla, or other flavouring.

Method. — Beat the yolks one by one into the arrowroot, and add flavouring to taste.

Beat the whites up stiffly, and stir them in lightly.

Pour the mixture into a greased pie-dish.

Bake for a few minutes, and serve as quickly as possible.

Treacle Posset.

- *Ingredients* — ½ pint of milk.
- ¼ pint of treacle.

Method. — Put the milk into a saucepan on the fire to boil.

When boiling, pour in the treacle.

This will curdle the milk.

Let it boil up again, and then strain it.

White-wine Whey.

- *Ingredients* — ½ pint of milk.
- 1 wineglass of sherry.
- Sugar to taste.

Method. — The same as in foregoing recipe. Sweeten to taste.

Orangeade.

- *Ingredients* — 2 oranges.
- 1 pint of boiling water.
- 3 lumps of sugar.

Method. — Take the rind thinly from half an orange.

Put it into a jug.

Peel the oranges, and slice them, removing the pips.

Put them into the jug.

Pour the boiling water over, add the sugar, and cover closely until cold; then strain.

273

Toast and Water.

- *Ingredients* — Toasted bread.
- Water.

Method. — Toast a piece of crust of bread nicely, being careful not to burn it.

Plunge it into a jug of cold water, and let it stand for thirty minutes.

Then strain the water from it.

Sago Gruel.

- *Ingredients* — ½ oz. of sago.
- ½ pint of water.
- 2 lumps of sugar.

Method. — Wash the sago, and let it soak in the water for thirty minutes.

Then simmer for about thirty minutes.

Add the sugar, and it is ready.

Prune Drink.

- *Ingredients* — 2½ oz. of prunes.
- 1 quart of water.

- 1 oz. of sugar.

Method. — Cut the prunes in two.

Boil them with the sugar in the water for one hour.

Strain, and cover until cold.

Rice Milk.

- *Ingredients* — 1 oz. of rice.
- 1 pint of milk.
- Sugar to taste.

Method. — Wash the rice, and simmer in the milk, with the sugar, for one hour.

Tapioca milk may be made in the same way. The crushed tapioca is the best.

274

Suet and Milk.

- *Ingredients* — 1 pint of milk.
- 1 oz. of suet.

Method. — Chop the suet finely.

Tie it loosely in muslin, and simmer in the milk for three-quarters of an hour; then strain.

Invalids' Soup.

- *Ingredients* — 1 pint of beef tea.

- 1 oz. of crushed tapioca, semolina, or sago.
- The yolks of 2 eggs.

Method. — Put the beef-tea into a saucepan on the fire.

When it boils, sprinkle in the tapioca; stir, and boil for about fifteen minutes.

Then add the yolks of the eggs; stir until they thicken, but do not let the soup boil after the yolks of the eggs are in it, as that would curdle them.

Gruel.

- *Ingredients* — 1 pint of water.
- 2 dessertspoonfuls of fine oatmeal.

Method. — Put the water on the fire to boil.

Mix the oatmeal smoothly with cold water.

When the water in the saucepan boils, pour in the oatmeal, and stir well until it thickens.

Then put it by the side of the fire, and stir occasionally, cooking it for *quite* half an hour.

Bran Tea.

- *Ingredients* — 3 tablespoonfuls of good bran.
- 1 quart of water.
- 1 oz. of gum arabic.
- 1 tablespoonful of honey.

275 *Method.* — Boil the bran in the water for ten minutes.

Dissolve the gum and honey in it, and strain it through muslin.

This is a remedy for hoarseness.

Linseed Tea.

- *Ingredients*—4 tablespoonfuls of linseed.
- 1 quart of boiling water.
- 6 lumps of sugar.
- 1 lemon.

Method.—Put the linseed and sugar into a jug, with the thin rind and juice of the lemon.

Pour boiling water over.

Let it stand, and then strain.

If the tea is preferred thick, two tablespoonfuls of the linseed may be boiled in the water.

Boiled Apple-water.

- *Ingredients*—3 good sized apples.
- 2 oz. of sugar.
- 1 quart of water.
- A little thin lemon-rind.

Method.—Wash the apples, and slice them.

Put them, with the sugar and lemon rind, into the water.

Boil gently for one hour.

Then strain, and cover close until cold.

Sole for an Invalid.

Grease a baking-sheet with butter.

Lay the sole on it.

Cover with greased kitchen paper, and put it into a moderate oven for fifteen or twenty minutes, according to the size of the sole.

If properly cooked, the sole will be as white and delicate as if it had been boiled.

276 It may be served with or without a plain white sauce.

Whiting, plaice, smelts, &c., may be cooked in the same way.

Chicken Fillets for an Invalid.

Cut some nice little fillets from the breast of a chicken, and cook them according to the directions in preceding recipe.

Sweetbreads plainly boiled.

Soak the sweetbreads in cold water for two hours.

Then put them in boiling water for six minutes.

Soak them again in cold water for twenty minutes.

Put them into boiling water or broth, and simmer them gently for thirty minutes or more, until quite tender.

Serve with or without a plain white sauce.

Other dishes suitable for the convalescent will found under the following headings:—

- Sole à la Béchamel.
- Sole à la Maître d'Hôtel.
- Whiting Boiled.
- Boiled Chicken.
- Sweetbread à la Béchamel.
- Mutton Chop.
- Rice Pudding.
- Cornflower Pudding.

- Blancmange.
- Tapioca Pudding.
- Sago Pudding.
- Haricot Soup.
- Tapioca Soup.
- Tapioca Cream.
- Oyster Soup.

277

SUPPER DISHES AND SALADS.

Ox Tongue.

Put it in lukewarm water; simmer for about three hours, until very tender. A very dry tongue may take four hours' gentle simmering. If very salt or much dried, soak for twelve hours before cooking.

When tender, remove the skin and cover with glaze or fine raspings.

Galantine of Fowl.

- *Ingredients* — 1 fowl.
- 1½ lb. of pork.
- 1½ lb. of veal.
- Yolks of 3 hard-boiled eggs.
- 2 truffles.

Method. — Bone the fowl, mince the pork and veal finely, and season with pepper and salt.

Fill the fowl with the stuffing, placing in the yolks and truffles.

Shape the fowl nicely, and fasten it securely in a cloth.

Boil it according to directions for boiling meat.

When cooked, remove the cloth and put in a clean one, fastening it as before.

Put it under pressure (not too much) until it is cold.

Remove the cloth, glaze it, and garnish with aspic jelly.

278

Galantine of Veal.

Breast of veal boned may be used instead of a fowl to make a galantine. Roll it round the stuffing and prepare it according to directions in preceding recipe.

Galantine of Turkey.

This may be prepared like Galantine of Fowl, using larger proportions for the stuffing.

Lobster Salad.

- *Ingredients*—1 fine lobster.
- 1 lettuce.
- 1 endive.
- 3 or 4 hard-boiled eggs.
- Some *mayonnaise* dressing.
- If possible, some aspic jelly.

Method.—Remove the flesh from the body and claws of the lobster, and cut it in pieces.

Let the lettuce be well washed and dried.

Cut it up, and mix it with the lobster and some *mayonnaise* sauce.

Put a border of chopped aspic on a dish.

Heap the salad in the middle.

Decorate the salad with pieces of endive and hard-boiled eggs cut in quarters.

Miroton of Lobster.

- *Ingredients*—A lobster.

- 1 lettuce.
- A small cupful of *mayonnaise* sauce.
- 6 hard-boiled eggs.
- If possible, some aspic jelly.
- Endive.

Method.—Cut the eggs at the bottom so that they will stand upright.

Then cut them in quarters, lengthwise.

279 Dip the ends in a little aspic jelly, or melted gelatine, and place them close together, in the form of a large circle on a flat dish with the white part inside.

Remove the flesh from the body and claws of the lobster.

Cut up the lettuce, and mix it with the lobster and *mayonnaise*.

Heap the salad in the middle of the crown of eggs.

Decorate it with endive, and put a border of aspic jelly round it.

Chicken Salad.

- *Ingredients*—A cold chicken.
- Some celery.
- A lettuce.
- Endive.
- Beetroot.
- A small cupful of *mayonnaise* sauce.
- 2 or 3 hard-boiled eggs.

Method.—Remove the skin of the chicken, and cut it into dice.

Cut up the celery into half-inch lengths, taking half as much celery as chicken.

Cut up the lettuce, and mix the chicken, celery, and lettuce together with the *mayonnaise*.

Put them into a salad-bowl, or heap on a dish.

Decorate with endive, beetroot, and hard-boiled eggs.

Mayonnaise of Salmon.

- *Ingredients*—Some cold dressed salmon.
- A lettuce.
- Endive.
- Some hard-boiled eggs.
- A small cupful of *mayonnaise* sauce.
- Some chopped aspic.

Method.—Break the salmon into flakes, removing the bones.

280 Cut up the lettuce, and mix the salad with the *mayonnaise* sauce.

Heap it lightly on a dish.

Decorate prettily with endive, and put some hard-boiled eggs, cut into quarters, round it; also, if liked, a border of aspic jelly.

Oyster Salad.

- *Ingredients*—1 tin of oysters.
- 1 crisp lettuce.
- 1 head of celery.
- A little *mayonnaise* or salad-dressing.

Method.—Wash the lettuce, and cut it coarsely.

Wash, and cut the celery into one-inch lengths,

Trim the oysters, and mix them with the salad.

Put the mixture into a salad-bowl, and pour over the *mayonnaise* or dressing.

Celery Salad.

- *Ingredients* — 2 heads of celery.
- 1 beetroot.
- A plain salad-dressing.

Method. — Wash the celery, and cut it into half-inch lengths.

Put them in a salad-bowl, and pour the dressing over.

Garnish with a border of beetroot.

Tomato Salad.

- *Ingredients* — A few ripe tomatoes.
- Equal quantities of oil and vinegar.
- 1 dessertspoonful of chopped parsley.
- Pepper and salt.

Method. — Slice the tomatoes and lay them on a glass dish.

Sprinkle them with the parsley.

Mix the oil and vinegar with pepper and salt, and pour over them.

281

Cauliflower Salad.

- *Ingredients* — 1 boiled cauliflower.
- A little *mayonnaise* or salad-dressing.
- Pepper and salt.

Method. — Divide the cauliflower into tufts, and remove the green leaves.

Place them on a dish, and pour the dressing over them.

Garnish with beetroot.

Potato Salad.

- *Ingredients* — Some boiled potatoes.
- 1 boiled onion.
- Some plain salad-dressing.

Method. — Slice the potatoes and onion thinly.

Lay them on a dish, and pour the dressing over.

If preferred, the onion may be omitted.

Haricot Salad.

- *Ingredients* — Some nicely cooked haricot beans.
- 1 teaspoonful of finely-chopped parsley.
- Equal quantities of oil and vinegar.
- Pepper and salt.

Method. — Lay the beans in a dish.

Sprinkle them with the parsley.

Mix the oil and vinegar with the pepper and salt, and pour over them.

Lentil Salad.

- *Ingredients* — Some boiled lentils.

- A little chopped parsley.
- Equal quantities of oil and vinegar.
- Pepper and salt.

Method. — Lay the lentils in a dish.

Sprinkle them with the chopped parsley.

282 Mix the oil and vinegar with the pepper and salt, and pour over them.

Mixed Salad.

- *Ingredients* — Equal quantities of boiled potato, carrot, turnip, and beetroot.
- Equal quantities of oil and vinegar.
- Pepper and salt to taste.

Method. — Cut the vegetables into small dice.

Place them in a salad bowl.

Mix the oil and vinegar with the pepper and salt, and pour over them.

Spring Salad.

- *Ingredients* — 1 lettuce.
- Some mustard and cress.
- Endive.
- Hard-boiled eggs.
- Beetroot.
- Watercress.
- Some *mayonnaise* or salad-dressing.

Method.—Wash the vegetables well; put them in a draught to dry them quickly.

Then cut them rather coarsely.

Put them into a salad-bowl.

Pour over the dressing, and garnish with hard-boiled eggs and beetroot.

For a more elaborate salad, put the vegetables into a glass or silver dish, heaping them high in the centre.

Decorate with sprigs of endive, placing a large tuft at the top.

Round the base place the hard-boiled eggs, cut in quarters, alternately with slices of beetroot.

Finish off with a border of chopped aspic jelly.

283

MISCELLANEOUS DISHES.

Cheese Pâtés.

- *Ingredients*—Some stale bread.
- ½ tablespoonful of hot water.
- 4 tablespoonfuls of grated cheese.
- 1 oz. of butter.
- A few bread-crumbs.
- Pepper and salt.
- A little cayenne.
- A few browned bread-crumbs.
- The yolk of an egg.

Method.—Cut the bread in slices of one inch in thickness.

Stamp into rounds with a circular pastry-cutter; scoop out the inside, making little nests of them.

Fry in hot fat (*see* French Frying); drain them on kitchen paper.

Put them inside the oven to keep hot.

Put the butter and water into a saucepan on the fire to boil.

When boiling, stir in sufficient crumbs to make the mixture stiff.

Beat in the yolk, add pepper, salt, and cayenne; and stir in the cheese.

Pile the mixture on the cases; sprinkle a few browned crumbs over them and be careful to serve quite hot.

284

Welsh Rare-bit.

- *Ingredients*—Some slices of bread about half an inch in thickness.
- Some slices of cheese.
- A little butter.
- The yolk of an egg.
- Pepper and salt.
- A little cayenne.

Method.—Toast the bread and keep it quite hot.

Cut the cheese into very thin pieces.

Put it in a saucepan with the butter; pepper and salt to taste.

Stir until it has melted, then mix in the yolk.

Spread it on the toast, and brown before the fire.

Toasted Cheese.

- *Ingredients*—Some slices of very hot toast.
- Some slices of cheese.
- Mustard, pepper and salt.

Method.—Toast the cheese nicely, and lay it quickly on hot toast.

Spread a little mustard thinly over it, with pepper and salt, and serve very hot.

Cheese Pudding.

- *Ingredients*—3 oz. of bread-crumbs.
- 1 pint of milk.
- ¼ lb. of grated cheese.

- 3 eggs.
- 1 oz. of butter.
- Pepper and salt.
- A little cayenne.

Method.—Put the crumbs into a basin.

Boil the milk; pour it over them, and let them soak.

285 Then add the yolks of the three eggs, the grated cheese, and seasoning.

Beat the whites of the eggs to a stiff froth and stir them in lightly.

Pour the mixture into a greased pie-dish, and bake in a quick oven until well thrown up and brown.

Macaroni and Cheese.

- *Ingredients*—¼ lb. of macaroni.
- 2 oz. of grated cheese.
- ½ pint of milk.
- 1 oz. of butter.
- ½ oz. of flour.
- Pepper and salt.
- A little cayenne.

Method.—Break the macaroni into small pieces, and boil in a quart of water for thirty minutes or more until the macaroni is tender.

Then strain away the water.

Melt the butter in a stewpan.

Mix in the flour smoothly.

Pour in the milk, stir, and boil well.

Then put in the macaroni, seasoning, and half the cheese.

Put the mixture into a greased pie-dish.

Sprinkle the remainder of the cheese over it, and bake in a quick oven until brown.

Macaroni Stewed in Milk.

- *Ingredients* — ¼ lb. of macaroni.
- 1 pint of milk.

Method. — Break the macaroni, and boil it in one quart of water for thirty minutes.

Then strain away the water, and pour in the milk.

Stew gently, stirring occasionally for thirty minutes.

This may be eaten with jam, sugar, treacle, stewed fruit, &c.

286

Macaroni Stewed in Stock.

Prepare according to directions in the preceding recipe, using stock instead of milk.

Macaroni is very good plainly boiled and served as a vegetable with roasted or stewed meat.

Savoury Rice.

- *Ingredients* — 1 onion.
- 2 oz. of rice.
- 1 pint of boilings from meat.
- Pepper and salt.

Method. — Boil the onion until tender, then chop it finely.

Wash the rice, and boil it in the meat liquor with the chopped onion until tender.

Add pepper and salt to taste.

Cheese Sandwiches.

- *Ingredients*—¼ lb. of grated cheese.
- The yolks of 3 hard-boiled eggs.
- 4 slices of buttered bread.
- 1 oz. of butter.
- Pepper and salt.
- A little cayenne.

Method.—Beat the yolks well with the butter; add the cheese and seasoning. Spread the mixture on the two pieces of buttered bread, and place the others over.

Rice Stewed with Cheese.

- *Ingredients*—½ lb. of rice.
- 2½ pints of water.
- 1 pint of milk.
- 2 oz. 287 of grated cheese.
- Pepper and salt.

Method.—Boil the rice gently in the water for half an hour, then add the milk and cheese and boil gently for half an hour more.

Stewed Normandy Pippins.

- *Ingredients* —1 lb. of pippins.
- 1 quart of water.

- 6 oz. of lump sugar.

Method. — Soak the pippins in the water.

Then stew them with the sugar for one hour or more until quite soft.

Place them on a glass dish and pour the syrup over them.

288

ODDS AND ENDS.

Croutons of Bread for Soup.

Cut stale bread into small dice, fry them in a little butter, or in a large quantity of fat (*see* French Frying), a golden brown colour. Drain on kitchen paper and serve on a folded napkin.

Toasted Bread for Soup.

Cut toasted bread into small dice, put them on a baking-tin and place them in a quick oven for a few minutes. Serve on a folded napkin.

Bread-crumbs.

These are best made by rubbing stale bread through a wire sieve, or the crumb of stale bread may be dried in a slow oven and pounded for crumbs.

Browned Bread-crumbs.

These can be made from white crumbs, which should be put on a baking-tin and baked a golden brown colour in the oven; or the crusts of stale bread can be dried in a slow oven and pounded. Raspings can be used, but they should be rubbed through a wire sieve.

Browned Crumbs for Game.

Put white crumbs into a frying-pan with a little butter, and stir until they are lightly browned.

Macédoine of Vegetables.

Cut carrots and turnips into fancy shapes with a dry cutter, boil them separately, cooking the turnips five minutes and the carrots fifteen. Mix them with nicely boiled green peas and French beans. In the winter Moir's *Macédoine* of Cooked Vegetables, sold in tins, will be found very convenient.

Pickle for Meat.

- *Ingredients* — 1½ lb. of salt.
- 6 oz. of brown sugar.
- 1 oz. of saltpetre.
- 1 gallon of water.

Method. — Put the salt, sugar, and saltpetre into a large saucepan with the water.

Put it on the fire, bring it to the boil, and let it boil for five minutes.

It must be kept well skimmed.

Strain it into a large tub or basin.

When the pickle is quite cold, meat can be put into it.

Fried Parsley.

Choose nice green parsley, wash and dry it, and pick it from the stalk; put it into a wire spoon or basket, and fry in hot fat (*see* French Frying). It must be removed directly it is crisp or it will discolour; drain it on kitchen paper, and sprinkle it with salt. Parsley that has been frozen will turn black in frying.

Rendering down Fat.

- *Ingredients* — 4 lb. of any fat, cooked or uncooked.

Method. — Cut the fat into small pieces.

Put it into a large saucepan and cover with water.

Boil for one hour with the lid on the saucepan, that the steam may whiten the fat.

290 Then remove the lid, and boil steadily until the water has evaporated, and the fat melted out of the pieces.

Stir occasionally to prevent the fat sticking to the bottom of the saucepan and burning.

When the fat is ready, let it cool a little, and then strain it.

The pieces should be well pressed to squeeze out all the fat.

This fat may be used for frying, or plain cakes and pastry.

The quantity given is sufficient for French Frying.

To clarify Dripping.

Melt the dripping and pour it into cold water.

When cold, scrape off the sediment which will be found at the bottom.

To clarify Butter.

Boil the butter, and remove the curd as it rises.

To blanch Almonds and Pistachio Kernels.

Put them into cold water, and bring it to boiling point.

Then remove their skins.

Almonds should afterwards be thrown into cold water to preserve their colour.

291

HOW TO USE UP FRAGMENTS.

Scraps of Bread.

These may be used for puddings, or dried and powdered for crumbs; they can also be used to thicken soup.

Cold Potatoes.

These may be mashed and baked in a pie-dish, or made into balls and fried or baked; they may also be sliced and made into French salad, or used to thicken soup.

Scraps of Meat.

If there are not sufficient to re-cook for a made dish of any kind, put them into the stock-pot.

Fat, cooked or uncooked.

This can be cut in pieces and rendered down (*see* Rendering down Fat). It can be used for frying, plain pastry, and cakes.

Fat Skimmings from the Stock-pot.

This is excellent to fry cutlets, &c., in, and can be used instead of butter.

Dripping.

Clarify it and use it for frying, plain cakes, and pastry.

292

Scraps of Cheese.

Grate them, and use for Welsh rare-bit, macaroni cheese, cheese sandwiches, *pâtés*, &c.

Cold Vegetables.

If any quantity, re-warm them, or make into French salads. Any scraps can be put into the stock-pot.

Water in which Vegetables have been boiled.

Use this, if possible, for vegetable soups, as it contains to a great extent the valuable salts of the vegetables.

Boilings from Meats.

These, if not too salt, can be used to make pea, lentil, and other vegetable soups.

293

FORCEMEATS.

Veal Stuffing.

- *Ingredients* — 3 tablespoonfuls of bread-crumbs.
- 1 tablespoonful of finely-chopped suet.
- 1 dessertspoonful of finely-chopped parsley.
- 1 teaspoonful of dried and powdered thyme and marjoram.
- 1 egg.
- Pepper and salt.

Method. — Mix all the ingredients with the egg well beaten.

A little grated lemon rind and juice improves the flavour.

Sage-and-Onion Stuffing.

- *Ingredients* — 4 onions.
- ¼ lb. of bread-crumbs.
- 7 sage leaves.
- 1 oz. of butter.
- Pepper and salt.

Method. — Blanch the onions by putting them into cold water, and bringing it to the boil; boil for five minutes, and then throw the water away.

Rinse the onions and put them into another saucepan of water, and boil for about one hour until they are quite tender; five minutes before taking them up put in the sage leaves.

Drain the onions and sage leaves, and chop them finely; then mix them with the bread-crumbs, pepper and salt.

Quenelle Forcemeat.

See Quenelles of Veal.

Forcemeat Balls.

These are made with veal stuffing. Shape it into balls and bake them in the oven. If they are served with hare, the liver is chopped and mixed with the forcemeat.

Imitation Foie Gras.

- *Ingredients* — ½ lb. of calf's liver.
- ¼ lb. of bacon.
- A piece of carrot, turnip, and onion.
- A sprig of parsley, thyme, and marjoram.
- A bay leaf.
- Pepper and salt.

Method. —Slice the liver, bacon, and vegetables.

Put them into a frying-pan and cook (turning frequently) until the liver is quite tender.

Care must be taken that the liver does not fry brown.

Put the whole contents of the frying-pan into a mortar and pound well. Then rub the mixture through a hair sieve.

PRESERVES.

Strawberry Jam.

- *Ingredients* — 8 lb. of strawberries.
- 4 lb. of loaf sugar.

Method. — Take the stalks from the strawberries and put them in a preserving pan.

Stir and boil for thirty minutes on a moderate fire.

Then add the sugar broken into small lumps; stir and boil for about thirty minutes longer, or until the jam stiffens.

Remove all the scum as it rises.

Put the jam into pots and cover close.

Raspberry Jam.

- *Ingredients* — 6 lb. of raspberries.
- 3 lb. of loaf sugar.

Method. — Remove the stalks from the raspberries and boil them over a moderate fire for fifteen minutes, stirring all the time.

Add the sugar broken into lumps, and boil for about thirty minutes longer, or until the jam will set.

Remove all the scum carefully.

Put the jam into pots and cover close.

Rhubarb Jam.

- *Ingredients* — 5 lb. of rhubarb.
- 5 lb. of lump sugar.

Method. — Peel and cut the rhubarb as for a tart, put it 296 in the pan with the sugar, and boil gently at first, then more quickly, skimming frequently.

When it will set it is ready.

Red Gooseberry Jam.

- *Ingredients* — 6 lb. of gooseberries.
- 3 lb. of lump sugar.
- Water.

Method. — Take the heads and stalks from the gooseberries and put them in a pan, allowing a quarter of a pint of water to every pound of gooseberries.

Put the gooseberries into a preserving-pan.

Stir and boil for fifteen minutes.

Then add the sugar.

Continue stirring until the jam is set, skimming frequently.

Put it into pots and cover close.

Damson Jam.

- *Ingredients* — 5 lb. of damsons.
- 3¾ lb. of lump sugar.

Method. — Boil for thirty minutes.

Then put in the sugar broken into small pieces, and boil and skim for about twenty minutes longer, or until the jam will set.

Put into pots and cover close.

Black-currant Jam.

- *Ingredients* — 5 lb. of black currants.
- 3¾ lb. of lump sugar.

Method. — Boil the fruit and sugar together until the jam will set, skimming all the time.

Put into pots and cover close.

297

MENUS.

I.

- Haricot soup.
- Boiled salmon, Hollandaise sauce.
- *Entrée.*
  Chicken croquettes.
- Saddle of lamb, mint sauce, spinach, potatoes.
- Cabinet pudding, orange jelly.
- Cheese, &c.
- Dessert.

II.

- Mock-hare soup.
- Boiled cod, egg sauce.
- *Entrée.*
  Curried rabbit.
- Roast leg of mutton, currant jelly, cauliflower, potatoes.
- Marmalade pudding, general satisfaction.
- Cheese.
- Dessert.

298

III.

- Celery soup.
- Boiled mackerel, melted butter.
- *Entrée.*
  Curried chicken.

- Boiled leg of mutton, caper sauce, mashed turnips, potatoes.
- Ginger pudding, apple turnovers.
- Cheese, &c.
- Dessert.

IV.

- Tapioca soup.
- Sole au gratin.
- *Entrée.*
  Mushrooms and kidneys.
- Roast fowl, bacon, bread sauce, Brussels sprouts, potatoes.
- Pancakes, snow pudding, cheese cakes.
- Cheese, &c.
- Dessert.

299

V.

- Macaroni soup.
- Fried cutlets of cod, anchovy sauce.
- *Entrée.*
  Minced meat with poached eggs.
- Braised breast of veal, cauliflower, potatoes.
- Marlborough pudding, jam roly-poly.
- Cheese, &c.
- Dessert.

VI.

- Lentil soup.
- Boiled cod, egg sauce.
- *Entrée.*
  Tomatoes stuffed with sausage-meat.
- Ribs of beef, horse-radish sauce, potatoes, spinach.
- Apple fritters, lemon pudding.
- Cheese, &c.
- Dessert.

300

VII.

- Haricot soup.
- Plaice filleted and fried, anchovy sauce.
- *Entrée.*
  Croustards of minced meat.
- Roast leg of mutton, red-currant jelly, potatoes, Brussels sprouts. Boiled fowl, egg sauce.
- Baked custard, sultana pudding, Normandy pippins.
- Cheese.
- Dessert.

VIII.

- Celery soup.
- Boiled halibut, shrimp sauce.
- *Entrée.*
  Rissoles.
- Sirloin of beef, horse-radish sauce, greens, potatoes.

- Tapioca pudding, jam tarts, raspberry pudding.
- Cheese.
- Dessert.

301

IX.

- Palestine soup.
- Fried whiting, thick white sauce.
- *Entrée.*
  Curried eggs.
- Shoulder of mutton, onion sauce, cauliflower, potatoes.
  Boiled beef, young carrots.
- Blancmange, gooseberry fool, apple pudding.
- Cheese.
- Dessert.

X.

- Mock-turtle soup.
- Boiled cod, lobster sauce.
- *Entrées.*
  Mutton cutlets à la macédoine. Tomato farni.
- Roast fillet of veal, boiled fowl, Béchamel sauce, asparagus, potatoes.
- Ducklings, green peas.
- Cheese cakes, chartreuse de fruit, lemon jelly.
- Cheese.
- Dessert.

XI.

- Clear soup.
- Cod's head and shoulders, oyster sauce. Fried smelts.
- *Entrées.*
  Beef olives. Quenelles of veal.
- Saddle of mutton, red currant jelly, spinach, potatoes.
  Boiled turkey, celery sauce.
- Grouse.
- Plum pudding, mince pies, tipsy cake, stone cream, cheese
  ramequins.
- Cheese.
- Dessert.

XII.

- Bonne femme soup.
- Boiled brill, anchovy sauce.
- *Entrées.*
  Podovies. Veal cutlets à la Talleyrand.
- Boiled leg of mutton, caper sauce, young carrots. Roast
  fowl and bacon, Brussels sprouts, potatoes.
- Goslings, green peas.
- Charlotte Russe, Viennoise pudding, apple fritters.
- Cheese.
- Dessert.

XIII.

- Calves'-tail soup. Palestine soup.
- Sole à la Rouennaise. Fried whiting.
- *Entrées.*
  Mutton cutlets, tomato sauce. Rabbit à la Tartare.
- Sirloin of beef, horse-radish sauce, spinach, potatoes. Boiled turkey, white sauce, Brussels sprouts.
- Pheasants.
- Marmalade pudding, general satisfaction, almond cakes, vanilla cream, cheese straws.
- Cheese.
- Dessert.

XIV.

- Mulligatawny.
- Cod, oyster sauce. Red mullets in cases.
- *Entrées.*
  Chicken tartlets. Fillets de bœuf, à la Béarnaise. Mutton cutlets à la macédoine.
- Saddle of mutton, red-currant jelly, potatoes, Brussels sprouts. Boiled turkey, celery sauce. Boiled tongue.
- 304 Pheasants.
- Apple amber pudding, plum pudding, stone cream, orange fritters, cheese ramequins.
- Cheese.
- Dessert.

XV.

- Mock turtle. Clear soup.

- Turbot, Hollandaise sauce. Lobster cutlets.
- *Entrées.*
  Braised sweetbreads. Pigeons à l'Italienne. Fillets of chicken.
- Saddle of lamb, mint sauce, asparagus, potatoes. Boiled fowls, bacon. Béchamel sauce, potato croquettes.
- Ducklings, green peas.
- Strawberry cream, Genoise pastry, cold cabinet pudding, claret jelly, cheese straws.
- Cheese.
- Dessert.

305

XVI.

- Potage à l'Américaine.
- Boiled turbot, lobster sauce.
- *Entrées.*
  Oyster patties. Fillets de bœuf à la Béarnaise.
- Roast leg of mutton, red-currant jelly, Brussels sprouts potatoes. Boiled fowl, bacon, celery sauce.
- Jugged hare.
- Gâteau de cerise, croquant of oranges, boiled custards.
- Cheese.
- Dessert.

XVII.

- Mock-turtle soup. Potato purée.
- Salmon, Hollandaise sauce, cucumber. Sole à la Béchamel.
- *Entrées.*
  Chicken à la Marengo. Braised sweetbreads.

- Saddle of lamb, mint sauce, peas, potatoes. Boiled fowl, egg sauce, boiled ham, potato croquettes, asparagus.
- 306 Goslings, peas.
- Strawberry charlotte, good trifle, orange jelly, jam puffs, cheese d'Artois.
- Cheese.
- Dessert.

XVIII.

- Julienne. Oyster soup.
- Turbot, lobster sauce. Sole à la Genoise.
- *Entrées.*
  Sweetbreads à la Béchamel. Mutton cutlets à la Rachel.
- Sirloin of beef, horse-radish sauce, asparagus, potatoes. Boiled leg of mutton, caper sauce, mashed turnips.
- Ducklings, peas.
- Charlotte Russe, pine-apple jelly, Normandy pippins, custards in glasses.
- Cheese.
- Dessert.

XIX.

- Ox-tail soup. Tapioca cream.
- Boiled salmon, Tartare sauce. Sole à la maître d'hôtel.
- 307 *Entrées.*
  Chicken à la Marengo. Mutton cutlets à la Milanaise. Podovies.
- Roast fillet of veal, French beans, potatoes. Haunch of mutton, red currant jelly.
- Goslings, green peas.

- Pistachio cream, orange jelly, snow puddings, cheese d'Artois.
- Cheese.
- Dessert.

308

SUPPERS.

Cold Supper. 12 People.

- One sirloin of beef.
- One roast turkey.
- One boiled ham.
- One lobster salad.
- One apple tart.
- Twelve cheese cakes.
- One blancmange.
- One jelly.
- Fruit.
- Cheese.

Cold Supper. 12 People.

- One rabbit pie.
- One galantine of veal.
- One ox tongue.
- One lobster salad.
- One charlotte Russe.
- One croquant of oranges.
- One small trifle.
- Two jellies.
- One dish of pastry.
- Cheese, &c.
- Fruit.

309

Cold Supper. 20 People.

- Ribs of beef rolled.
- Salmon coated with mayonnaise sauce.
- Cucumber.
- One pigeon pie.
- One veal-and-ham pie.
- One ox tongue.
- A stone cream.
- One tipsy cake.
- A dish of Genoise pastry.
- A pine-apple jelly.
- A compote of peaches.
- Strawberries and cream.
- A lemon jelly.
- Cheese and fruit.

Cold Supper. 20 People.

- Roast turkey.
- Boiled tongue.
- One pigeon pie.
- One mayonnaise of lobster garnished with aspic.
- One veal-and-ham pie.
- One large pine-apple jelly.
- One large blancmange.
- Jam puffs.
- Cheese cakes.
- Boiled custards.
- Normandy pippins.
- Gâteau de cerise.
- Cakes.
- Biscuits.
- Fruit.

High Tea for 12 People, for Lawn Tennis Parties, &c.

- One lobster mayonnaise.
- Two chickens, coated with thick white sauce.
- One veal-and-ham pie.
- One tongue.
- One apple amber pudding.
- Twelve boiled custards.
- Twelve custard-glasses filled with chopped jelly.
- One blancmange.
- Strawberries and cream.
- Thin bread and butter.
- Biscuits, cake, fruit.
- Tea, coffee.

Supper for 50 People. Guests not seated.

- Ham, tongue, and beef sandwiches.
- Four quarts of blancmange, differently flavoured and decorated.
- Four quarts of jelly, differently flavoured and moulded.
- Two charlottes Russe.
- One large trifle.
- Two tipsy cakes.
- Three dishes of Genoise pastry in various forms.
- Two lemon sponges.
- Fruit.

Supper for 50 People. Guests not seated. Less expensive.

- Ham and beef sandwiches.
- Four quarts of blancmange, differently flavoured and decorated.
- Two quarts of apple gâteau.
- Four jellies of different kinds.
- One large lemon-sponge.
- Two dishes of light pastry.
- Boiled custards.
- Normandy pippins.
- Two jaunemanges.
- Two tipsy cakes.
- Fruit.

313